Valley of Forgetting

Valley of Forgetting

*Alzheimer's Families
and the Search for a Cure*

JENNIE ERIN SMITH

RIVERHEAD BOOKS | NEW YORK | 2025

RIVERHEAD BOOKS
An imprint of Penguin Random House LLC
1745 Broadway, New York, NY 10019
penguinrandomhouse.com

Copyright © 2025 by Jennie Erin Smith

Penguin Random House values and supports copyright. Copyright fuels creativity, encourages diverse voices, promotes free speech, and creates a vibrant culture. Thank you for buying an authorized edition of this book and for complying with copyright laws by not reproducing, scanning, or distributing any part of it in any form without permission. You are supporting writers and allowing Penguin Random House to continue to publish books for every reader. Please note that no part of this book may be used or reproduced in any manner for the purpose of training artificial intelligence technologies or systems.

Riverhead and the R colophon are registered trademarks of Penguin Random House LLC.

Book design by Cassandra Garruzzo Mueller

LIBRARY OF CONGRESS CATALOGING-IN-PUBLICATION DATA

Names: Smith, Jennie Erin, 1973– author.
Title: Valley of Forgetting : Alzheimer's Families and the Search for a Cure / Jennie Erin Smith.
Description: New York : Riverhead Books, [2025] | Includes bibliographical references and index.
Identifiers: LCCN 2024044615 (print) | LCCN 2024044616 (ebook) | ISBN 9780525536079 (hardcover) | ISBN 9780525536093 (ebook)
Subjects: MESH: Restrepo Lopera, Francisco, 1951–2024. | Alzheimer Disease | Family Support—history | Biomedical Research—history | Research Subjects—psychology | Colombia
Classification: LCC RC523 (print) | LCC RC523 (ebook) | NLM WT 155 | DDC 362.1968/311—dc23/eng/20241115
LC record available at https://lccn.loc.gov/2024044615
LC ebook record available at https://lccn.loc.gov/2024044616

Printed in the United States of America
1st Printing

The authorized representative in the EU for product safety and compliance is Penguin Random House Ireland, Morrison Chambers, 32 Nassau Street, Dublin D02 YH68, Ireland, https://eu-contact.penguin.ie.

For the families

CONTENTS

Part 1
Curious Doctors
1

Part 2
Las Familias
117

Part 3
Resilience
231

ACKNOWLEDGMENTS 349

NOTES 355

INDEX 367

Part 1

Curious Doctors

ONE

A thick brown file tells the story of a man named Pedro Julio Pulgarín, who on October 9, 1984, was escorted against his will through the gates of the San Vicente de Paúl hospital in Medellín, Colombia.

Pedro Julio was forty-nine years old, balding, and missing his upper teeth. He was what local people called a *montañero*—a term that loosely translates to "hillbilly," but with a nobler connotation. Pedro Julio had raised cows in the mountains above a town called Belmira for his entire adult life, until a few months before, when he'd started wandering off into pastures and getting lost. His family feared that, in that rugged terrain of hills and boulders, he would get himself killed. His wife and some of their children left their highland farm for a house in town, where they could secure him indoors and keep tabs on him. There, however, things only got worse. Pedro Julio grew frustrated, filled with the urge to wander, and his family grew concerned that he might get hold of a knife or rock and hurt them if they didn't let him out.

This was out of character for Pedro Julio, who was known to all as a loving father, a Catholic devoted to his saints, a hard worker, and shrewd. He had saved money furiously, acquiring enough through leases and purchases and fortunate bets in country taverns to leave some land to every one of his twelve living children. Now he could not remember their names. A rural doctor had treated Pedro Julio with niacin to see if his confusion might improve, but the vitamin had no effect, and the doctor advised the family to take him to a hospital in Medellín, a three-hour journey over unpaved roads.

The San Vicente hospital was affiliated with the University of Antioquia, Medellín's preeminent public university. Its early twentieth-century campus of low buildings with tall windows, tile roofs, chapels, and gardens was a relic from an era when hospitals were meant to heal souls as well as bodies. A neurology resident named William Cornejo, one of the first doctors to evaluate Pedro Julio, entered his notes by hand:

> *Farmer, two years of primary school and worked until two months ago. Until four years ago he was completely fine; he was intelligent. Two years following onset he could not distinguish family relations. Four months ago he presented with: marked disorientation in general, and in space and time. Neglect of personal hygiene. Emotional laity, passing easily from crying to laughing, with delusions of persecution; he hears and sees things. Irritability and incoherence. At the initiation of his current illness he complained of pain in the back of his head.*

In the clinic Pedro Julio barely spoke, except in the empty phrases people used to fill space in conversation: *Eavemaría,* Lord almighty; *Así es la cosa,* here's the thing. He could still make the sign of the cross, but he could no longer count more than two digits in a row. He lost his patience with Cornejo, saying he'd have no more *pendejadas,* bullshit.

Three of Pedro Julio's uncles had died after a similar course of illness, his wife reported; so had a brother, at age fifty-two. His mother had died in a government mental hospital at fifty, apparently of the same causes.

"If we'd been practicing medicine then the way we're supposed to today, I would have prescribed him a sedative for the agitation and sent him home," Cornejo recalled. Instead Cornejo hospitalized Pedro Julio. Here was what looked like a case of senile dementia, in a man too young to have it. *Alzheimer's?* he wrote in his notes.

The next day, however, Pedro Julio left the hospital. He walked out through the front gates of San Vicente and into the streets of downtown Medellín as though they were the green, boulder-strewn hills of Belmira. *Patient escaped the service*, his file read. *The patient took leave despite being under our observation and was followed. It was impossible to make him return as he became aggressive. The guards and doorman could not detain him. The social workers and family have been advised. He was headed downtown.*

More than three decades later, William Cornejo still remembered the fright Pedro Julio's disappearance gave him. When the patient was returned to the hospital, late that same day, the resident sought help in making sense of this wandering farmer.

He turned to his colleague Francisco Lopera Restrepo.

Handsome and unflappable in his demeanor, with wavy hair already turning gray, Lopera was a third-year neurology resident whom everyone called Pacho. "Pacho was really into neuropsychology," Cornejo recalled. "He'd been studying how language worked, and memory, and all that."

At thirty-three, Lopera was a little older than the other residents. It had taken him nine years to finish his undergraduate degree, thanks to protests and strikes, and after medical school he'd put in a three-year stint on the country's Caribbean coast, near Panama, as a government physician.

VALLEY OF FORGETTING

LIKE HIS PATIENT Pedro Julio Pulgarín, Lopera was a *paisa*, a term that usually means "compatriot" or "countryman" but in Colombia refers to anyone with roots in Antioquia—the mountainous central-northwest department that is home to Medellín and a constellation of country towns. Paisas are famed as polite, thrifty, hardworking, and supremely religious. They will cross themselves in front of churches, even while passing them on the metro, and bless you while selling you potato chips on a bus. They maintain certain rural ways, even in the city. The traditional paisa diet consists of simple foods prepared without much spice: beans, meats, rice, hearty soups, arepas made of ground corn, fruit juices, and fresh cheese. Some paisas are still wary of the fork and knife, considering the spoon up to any challenge.

Paisas are the heirs of Spanish colonists who clambered through these mountains on mule trains in the sixteenth and seventeenth centuries, looking for gold. They are also descended from the Indigenous Colombians who first broke those mountain paths, and of the Africans forced into slave gangs to dam and dredge the gold-bearing rivers, but they tend to downplay these lineages in favor of their European roots. In the eighteenth century the paisas turned to agriculture, establishing fincas in Antioquia's cooler uplands. They built farmhouses of wood beams and earthen walls that they decorated with hanging plants and filled with children—such staggering numbers of children that some of the old houses feel more like convents or dormitories. They grew sugar and coffee and every type of fruit and vegetable in the fertile soils of Antioquia, and with centuries of experience moving goods around on mules, and of hiding their gold from the tax collectors of the Spanish crown, they proved good at business and finance. By the late nineteenth century, Medellín, the hub of their trade and the seat of their power, became so prosperous that envious elites in

Bogotá spread rumors that paisas were the descendants of Jews. It was meant as a slander, but the paisa establishment grew to embrace it, in part because it resonated with their ideas about being abler, more European, a *raza* apart.

Lopera was raised some twenty miles from where Pedro Julio lived, in the same Catholic, conservative, rugged mountain culture. The 1950s, when he was growing up, were the height of La Violencia, a time of political massacres all over Colombia. A reactionary local bishop, Miguel Ángel Builes, had loudly and repeatedly declared liberalism to be a mortal sin, and he encouraged locals to deal with the threat as they saw fit. Lopera's grandfather was killed in a knife fight, and as a little boy he watched his father, a conservative, stand in the town square and deliver an impassioned speech begging people to stop murdering liberals.

Lopera was the fourth of thirteen siblings. Between his mother's giant clan and his father's, he counted one hundred first cousins. At age thirty-three he still didn't know how to drive a car, though he was excellent on a horse.

After examining Pedro Julio, Lopera concurred with Cornejo that the farmer had symptoms of dementia, but its cause was a mystery. The doctors gave him drugs for parasites, checked his thyroid, tested for syphilis, and looked for copper and arsenic in his blood, trying to identify any potentially reversible cause. They stuck EEG probes all over his head and ordered a CT scan—a new technology to them, though already more than a decade in use elsewhere.

The scan revealed deep grooves in Pedro Julio's brain, showing that it was shrinking. The most dramatic shrinkage was in the temporal lobes, where speech is interpreted and new memories are encoded. The ventricles in the center appeared enlarged, another sign of atrophy. The residents thought about conducting a brain biopsy—the only way besides an autopsy to confirm the presence of the microscopic plaques and tangles of abnormal proteins considered the hallmarks of Alzheimer's disease—but this was terribly invasive, and ultimately Lopera and Cornejo decided against it. Pedro Julio was released to the care of his family, as there was little else the doctors could do.

THE CASE OF Pedro Julio nagged at Lopera: there was something under the surface, he felt sure, and the best way to understand it was to go to the scene. In December 1984, Lopera and Cornejo piled into Cornejo's Romanian-made car and traveled to the Pulgarín farmstead. The drive to Belmira is a steady ascent from Medellín, and Cornejo forced Lopera to get behind the wheel and learn to drive as they lurched upland, passing the small town of San Félix with its modest single-domed church and the bigger town of San Pedro with its grand triple-domed church. A few hours later they encountered a landscape of cool streams and grassy open slopes, like something you might find in Ireland or Scotland, with a Gothic-style cathedral at its center. The Pulgarín farm sat above Belmira in a cold, misty hamlet; to reach it, the doctors had to leave their car behind and borrow animals to finish the journey.

From what they had learned about Pedro Julio's mother and uncles, Lopera and Cornejo already suspected they were dealing with a hereditary disease. But an environmental cause was also possible. In the countryside there were many potential triggers of dementia, including chemicals used in mining and agriculture. At the time, aluminum buildup in the brain was still widely suspected as a cause of Alzheimer's disease. When Lopera and Cornejo arrived at the Pulgarín farm, they found that one of Pedro Julio's brothers, who was two years younger, was experiencing similar symptoms of forgetting and confusion; he had just left his gun in a tavern. The doctors put the brother through some pencil-and-paper tests for dementia, asking him to draw a house, a cross, a clock, a cube. It took him three tries to make a wobbly-looking cross, five to draw a cube.

Lopera and Cornejo only vaguely understood these tests or how to administer them. Nor did they know much about using genealogies to uncover a family history of inherited illness. Nevertheless, with the Pulgaríns gathered around them, they started building a chart, or genogram, of the Pulgarín family tree. Lopera drew squares and circles representing male and female members, with crosses to mark the deceased.

The exercise revealed that nine people in the family had developed this distinctive early-onset dementia over four generations. Only those who had a parent with the disease appeared to be at risk of developing it. This seemed to exclude a viral or chemical cause, suggesting that the disease was dependent on genetic factors. Dementia set in around forty-seven or forty-eight, on average, and patients died within ten years of its onset.

In the months that followed, Lopera and Cornejo returned to the Pulgarín farm several times: by bus; by car; in a university vehicle with Lucía Madrigal, a nurse from the neurology department. In 1987, they published the first paper about the family in a Colombian medical journal. Lopera and Cornejo had read accounts of other families with early-onset dementia—one in the American Midwest and another in Canada—and they cited these in the paper. The Pulgarín genealogy had revealed important parallels between the Colombian family and the North American ones: Every sick person had had a sick parent, meaning that the disease did not skip generations. It occurred in both men and women. All this pointed to a disease caused by an autosomal dominant mutation—an abnormal gene that can be passed from parent to child regardless of sex. Each person born to a parent carrying an autosomal dominant mutation has a 50 percent chance of inheriting it. If someone is born without it, the chain is broken, and his or her descendants are unaffected.

In that era, researchers were excited to discover families with early-onset Alzheimer's. Such families were beginning to be considered key to unlocking the disease's secrets, both the genes setting it in motion and the nature of its course. Lopera and Cornejo, not wanting to waste a promising research opportunity, wrote at the end of their paper that they intended to conduct "a careful follow-up of the family with the aim of carrying out histopathology"—a subtle way of saying that they wanted the Pulgaríns' brains when they died. They strongly suspected the illness to be a form of Alzheimer's disease, although they could not make a definite diagnosis without putting brain tissue under a microscope. They settled on calling it "early-onset dementia of the Alzheimer's type," a conservative description. They planned, they wrote, to collect blood and fingerprints from the Pulgaríns, employ more sensitive cognitive tests to try to detect the disease earlier, and enlist the help of geneticists to begin untangling this fatal inheritance.

IN LATE 2017, I moved to Medellín to learn about the researchers who had discovered this famous cluster of early-onset Alzheimer's disease, the families who lived with it, and the science that was emerging from their decades of close collaboration. By then Lopera had built a whole research institution, the Grupo de Neurociencias de Antioquia, around his work with this kindred. Their unique genetic mutation, dubbed the paisa mutation, had been found to be as old as the Spanish conquest of the Americas. Lopera made no bones about his ambition: after decades of work, he was hoping to discover, in this cohort of rural Colombians and their children and grandchildren, a treatment, maybe even a cure, for Alzheimer's disease. The large extended family he studied counted some six thousand members, making it bigger on an order of magnitude than any other early-onset Alzheimer's cluster known to exist.

Now that Medellín had finally calmed down after decades of violence, foreign researchers had come to see in this family an ideal testing ground for drugs that might stop Alzheimer's disease before it could take hold. One pharmaceutical company had poured more than $100 million into a clinical trial at the Grupo de Neurociencias—or just Neurociencias, as the researchers called it—to test a drug called crenezumab in the paisa mutation families. Among the participants Lopera and his team enrolled in this trial were the children and grandchildren of Pedro Julio Pulgarín and his siblings, people he had first come to know as a young doctor and had painstakingly followed for generations.

In Gabriel García Márquez's novel *One Hundred Years of Solitude*, the traveling gypsy Melquíades cures the residents of Macondo, a fictional town in northern Colombia, of a mysterious disease causing them to lose their memories and their sleep. García Márquez was not aware, at the time he was writing, of the families with Alzheimer's disease in the mountains of Antioquia, but the similarities were striking, and the investigators alluded to them in interviews with newspapers and TV stations. Would crenezumab prove to be, as they dearly hoped, like the magic elixir of Melquíades? It would be several years before anyone knew, and expectations remained high.

The crenezumab trial was just the largest and most expensive of all the studies going on at Neurociencias. With such a large early-onset Alzheimer's population, the number and type of studies you could design—whether in imaging, neuropsychology, pathology, or cell biology—were almost infinite.

Lopera and his colleagues, who had long believed the paisa mutation family to be the unique product of Antioquia's supposed geographical and racial isolation, were lately discovering more local families with previously unknown genetic mutations that caused early-onset Alzheimer's. Why were there so many in one small corner of the world? Perhaps more surprising still, rare outliers had appeared among the mutation carriers: a handful of people who possessed the devastating gene but got sick much later than they were expected to, and sometimes not at all.

ONLY PART OF THIS story belonged to the scientists. The rest belonged to *las familias*, as the investigators referred to their thousands of research participants. Since the 1990s, they had donated hundreds of their loved ones' brains to Neurociencias. They had subjected themselves to scans, spinal taps, and infusions of radioactive compounds. Now many were receiving injections of an experimental drug that might not even become available to them if it worked, given the vagaries of the global pharmaceutical trade and strained budgets for healthcare in Colombia.

Neurociencias hosted frequent visits from journalists and film crews. They came from Germany, Japan, the United States, or the United Kingdom seeking interviews with Francisco Lopera, the charismatic doctor with the near-constant smile and thick white mane, and with the families living in bucolic hamlets or humble city houses who nursed their sick beloved. The family members presented to the media tended to be people with only praise for the investigators and high hopes for a breakthrough. But this left a lot unsaid about how they shouldered the burden of their inheritance over months, years, generations. How did the specter of early Alzheimer's shape their lives and decisions? Were they as intent as the scientists on unlocking the mysteries of their disease? How did it feel to be part of a study that had enrolled their parents and grandparents, and would one day enroll their children? If the current clinical trial failed, would they despair?

The family members seldom knew whether they carried the paisa mutation, and the handful interviewed by the media often said they did not wish to—they preferred to trust in God. I wondered about this. I had lived in Colombia before, and knew that it was a country of faith, perhaps nowhere more than in Antioquia. But it was also a country of rising education levels and changing concepts of personal freedom. More people lived in cities than in rural areas; more women worked than stayed home. Fertility rates were only barely higher than in the United States, and even a longtime ban on abortion was being eroded. It stood to reason that there would be some people in this cohort of six thousand—young people especially—who would want to learn whether they carried a mutation that destined them to fall into dementia by their forties, to have their lives cut in half.

BY 2017, MEDELLÍN had become a magnet for foreign tourists, in part because of its powerful redemption story. Huge investments in public transit had given rise to tramways and gondolas linking the business districts to the sprawling hillside *comunas*, helping soften the old geographical boundaries separating rich and poor. Public service campaigns implored people to stop playing music loudly, to respect one another, to embrace "culture." The idea was to make life kinder and more civil for the residents, who had been through a lot. Its tree-lined canals and bike paths and immaculate public parks were testament to the city's civic pride, its constant efforts toward self-improvement.

Downtown, near the University of Antioquia and the San Vicente hospital where Lopera and Cornejo still practiced medicine, the city remained a little more defiant. Boys still wore rosaries around their necks, over the green jerseys of the Atlético Nacional soccer team, as they'd been doing since the days of Pablo Escobar. Street flyers advertised the services of witches: *Bring a photo, item of clothing or hair of your loved one and we 100% guarantee their return. In 3 days they will be at your feet begging your forgiveness. We also work farms, cattle, crops.* Porn vendors set up their tables against the wall of the seventeenth-century basilica, and old men sat with typewriters under umbrellas, waiting for someone who needed to dictate a letter or a form.

Beyond the hospital's main gate—the same gate through which Pedro Julio Pulgarín had escaped in 1984—is a wide boulevard lined with mortuaries and a large, beautiful church with a terra-cotta facade, used mainly for funerals. The view from Francisco Lopera's office, at the University of Antioquia's medical research complex, looked out over the church's roof.

Lopera's schedule was always full, and it seemed as though when he wasn't being interviewed he was being fêted—by the city council, the national government, by Colombian diplomats in Washington, D.C. Recently his hometown announced it was placing a plaque on the house where he was born.

His old colleague William Cornejo, meanwhile, was about to retire. I met Cornejo at the San Vicente hospital, where we walked by his and Lopera's old neurology department—a cramped first-floor space, long since converted into a laundry—and past the hospital's central fountain, shaded by royal palms, where nurses in stiff white uniforms and old-fashioned caps sat checking their phones. Orderlies rolled patients on gurneys with giant wheels out onto the wide patios for some sun, just as in the days of tuberculosis.

Cornejo was gray and slightly hunched, wearing a corduroy blazer. Except for a few years in France, he had worked here for his entire medical career: as a student, a resident, an investigator, a professor, and finally an administrator. Academic medicine was his life. It had been his refuge and oasis, he explained, during the years when drug mafias owned the city, when he left the hospital one afternoon to find armed thugs trying to steal his car—the same one he and Lopera had used to get to the Pulgarín farm. "I'm sure they wanted it for a car bomb," he said.

When Cornejo told me about his discovery of the first Alzheimer's case, he was the story's protagonist. It was true that Cornejo was the first physician to see Pedro Julio Pulgarín; it was his idea to hospitalize the farmer, and later to memorialize his case in a paper.

"But it was Pacho who really sniffed out its importance," Cornejo acknowledged, with what I took to be a twinge of regret.

THE ERA IN WHICH Lopera and Cornejo discovered the Pulgarín family and published their first paper was a dark one for Medellín. The city then was the turbulent center of a fast-growing cocaine trade dominated by Pablo Escobar, a schoolteacher's son who had cut his teeth stealing, grinding down, and reselling cemetery plaques on the streets by the San Vicente hospital. By the early 1980s, Escobar had become a household name; all of Colombia knew about his private zoo and his fleet of airplanes and the dead bodies in his wake. Escobar was even dabbling in politics, in the hope that holding elected office would protect him and his fellow capos from being extradited to the United States, but Colombia's establishment soured on them fast. Escobar's assassins murdered ministers, judges, journalists, investigators, and politicians who got in the cartel's way. The bonanza of cocaine money, and the lawless climate it fed, gave rise to a new breed of right-wing paramilitaries, who worked in tandem with the cartel to advance their own violent agendas. The paramilitaries targeted anyone they deemed a guerilla sympathizer, and that included professors and students at the University of Antioquia.

"We were very scared of what was happening," Lopera recalled, "but at the time we didn't have a complete sense of it. One thing after another would occur, and only later did we have a sense of the magnitude of the tragedy. It never would have occurred to us, for example, that Héctor Abad would be murdered."

Héctor Abad Gómez, a physician and a former professor of Lopera's, was Colombia's foremost public-health expert and a rising politician. In the 1950s, Abad had helped lobby for the creation of the *año rural*, or obligatory government service for physicians, insisting that recent graduates of medicine get out into the real world for a year. Lopera admired Abad. He "helped us understand the problems of the population," Lopera said, such as malnutrition, infant mortality, and poverty. In class Abad would embarrass his students, many of whom had patrician backgrounds, with questions they could not answer: "Lopera. What is the infant mortality rate in Colombia? Why don't you know this? If you plan to be a doctor you need to know how many children die in this country and of what." Abad was neither a dogmatic leftist nor a guerilla sympathizer. But his relentless advocacy for the health of the country's poor earned him the ire of local conservatives and their new paramilitary allies.

In 1987, the mutilated corpses of students and professors at the University of Antioquia began turning up around the city. A professor of dentistry was shot at his weekend home, and the remains of one of his students appeared days later. The body of a veterinary student was found on a roadside with his nose broken, fingers cut off and an eye burst.

At first it was uncertain whether the killings were related, until an audiotape appeared from a paramilitary group threatening to "scrub" the city of leftist influences. The University of Antioquia was a hotbed of left-wing political activity and even home to guerilla cells, but not all the dead had political ties. One student, shot eight times and dismembered, was targeted because he'd spent nearly a decade studying, taking one course at a time. This raised suspicions that he was a guerilla infiltrator, when in fact he'd dragged out his studies so he could work and support his eighty-two-year-old father, who now had to bury him.

In the summer of 1987, the paramilitaries breached the university campus. They shot an anthropology professor in the head while he sat chatting in a café. They firebombed a university bus. Classes were canceled for months as panic set in. In August, Héctor Abad organized a protest march, leading three thousand students, professors, and their supporters from the campus into downtown Medellín, holding red carnations and pleading for an end to the murders. The florists ran out of carnations to sell them.

Weeks later two men on a motorcycle shot Abad in the head and chest as he chatted with his friend Leonardo Betancur, another physician turned politician. When Betancur tried to escape into a building, the assailants caught and killed him, too. By December, seventeen students or professors at the University of Antioquia had been killed, among them a pharmacy student, a medical student, and four doctors, most of them active in left-wing politics. The last physician to die was an internist, killed by gunmen inside the San Vicente hospital.

Lopera was no stranger to the dangers of Colombia's armed groups; during his government medical service he had been kidnapped briefly by guerillas with the Revolutionary Armed Forces of Colombia, or FARC, and forced to treat a fighter who had been shot in the leg. He had survived that experience, but the university murders erased any lingering illusion of safety. Héctor Abad "was someone who defended the people, who protected them, advocated for them," Lopera said.

"And that's who they kill? It was painful for me and for everyone here and created an incredible panic because from that moment forward, we were all at risk and we knew it."

Lopera and his colleagues sought fellowships abroad, in any country that would have them, any discipline where they could be of help. Lopera applied for a post in Belgium, to study behavioral neurology in children and adults. He packed the hand-drawn Pulgarín genealogy among his belongings, not knowing what he'd do with it in Brussels. His colleague William Cornejo left for a fellowship in France.

When Pedro Julio Pulgarín died in Belmira, late that same year, there was no one around to collect his brain.

TWO

Alzheimer's disease was first described in a patient barely older than Pedro Julio Pulgarín. Auguste Deter was fifty-one when she was admitted to Frankfurt's Hospital for the Mentally Ill and Epileptics in 1901. Auguste, a housewife, had become jealous of her husband after a long and happy marriage. Her memory had rapidly deteriorated along with her speech, and at times she thought someone was trying to kill her.

Soon after she was admitted, the physician Alois Alzheimer came to see her. He spent four days evaluating Deter and making notes.

> She sits on the bed with a helpless expression. What is your name? *Auguste.* Last name? *Auguste.* What is your husband's name? *Auguste, I think.* Your husband? *Ah, my husband.* She looks as if she didn't understand the question.

In an era dominated by Sigmund Freud's psychoanalytical theories, Alzheimer belonged to a far less celebrated scientific school that focused on the organic, physiological roots of mental disorders. Though he worked in psychiatric hospitals for most of his career, he had been trained as an anatomist and distinguished himself as a pathologist, adept at finding abnormalities in the microscopic structures of brain tissues.

Alzheimer asked Auguste straightforward questions. He asked her what she was eating as she lunched on pork and cauliflower. Spinach, she told him. Her handwriting faltered, and though she could still read aloud with extreme care, she seemed not to understand the meaning of what she had just read. Alzheimer asked her to do simple math: If you buy six eggs, at seven dimes each, how much is it? Differently, she replied.

"I have lost myself," Auguste Deter told the doctor repeatedly. Alzheimer followed up on Auguste for four and a half years, checking in even after he'd moved from Frankfurt to a psychiatric hospital in Munich. In Auguste's original case file, which went missing for nearly a century before being rediscovered, a photograph shows her as a prematurely aged woman with deep facial wrinkles, folded hands, and stringy hair, her pale eyes unfocused. Auguste spent her final year unresponsive in a hospital bed, incontinent, with bedsores that gave rise to a fatal infection. When Auguste died, in 1906, Alzheimer made sure to get hold of her brain.

The next year he published his description of "a Peculiar Disease of the Cerebral Cortex." It was what Alzheimer found in Auguste's brain tissue, more than what he'd seen at her bedside, that set the disease apart to him. Her brain was atrophied, weighing much less than it should have, and something bizarre had happened to the neurons. Looking at slices of tissue stained with silver solutions, Alzheimer noticed that inside as many as a third of them, there were "very striking changes of the neurofibrils," the filaments that give nerve cells their structure. "Combined in thick bundles, they appear one by one at the surface of the cell," Alzheimer observed. The bundled fibrils reacted to stains differently from healthy ones, suggesting they had undergone a chemical change. "Finally," Alzheimer wrote, "the nucleus and the cell itself disintegrate and only a tangle of fibrils indicates the place where a neuron was previously located."

Alzheimer also saw, besides the mysterious tangles and dead neurons, that there were seed-shaped deposits of an unknown material—"a special substance," as he called it—dispersed throughout the cortex, or outermost layer of the brain. Alzheimer did not speculate as to how or whether the two types of lesions, the seed-shaped deposits and the neurofibrillary tangles, were related. But they struck him as altogether remarkable. "Considering everything, it seems we are dealing here with a special illness," he concluded, and he urged that his fellow clinicians look deeper into such dementias, including conducting careful autopsy studies.

Alzheimer went on to study a few more brains from people like Auguste Deter, but never lived to discern whether what he had seen in Auguste was a disease process unique to younger patients, or an unusually early and severe presentation of the same kind of dementia that was common in the old. In 1912, Alzheimer suffered a chest infection that damaged his heart and led to his death three years later. He was fifty-one, the same age as his most famous patient on the day she entered his world.

Alzheimer's disease as it was first defined—as a progressive, young-onset dementia accompanied by the presence of microscopic seedlike plaques and neurofibrillary tangles in the brain—was for decades considered very rare. Indeed, it was almost forgotten until the late 1960s, when autopsy studies began to show that half of elderly patients who had died with dementia had the same brain pathology that Alzheimer had described. By the 1980s the medical community had broadened the definition of Alzheimer's disease to encompass both a common late-onset form that occurred in people older than sixty-five, and a much rarer early-onset form that usually ran in families. These were deemed the same disease regardless of the patient's age, though some researchers thought there might be subtle differences between the early and late forms.

Francisco Lopera and William Cornejo, in their 1987 paper on the Pulgaríns, had identified some, but not all, of the relevant research on familial Alzheimer's disease. This was before the internet made papers easy to find, and scientists in Colombia, limited by resources and geography, would have had difficulty staying on top of the literature.

In 1984, the year Lopera and Cornejo encountered the Pulgarín family, a pair of California scientists, George Glenner and Caine Wong, reported discovering the core component of the plaques seen in Alzheimer's brains: a protein called amyloid-beta. Sometimes referred to as beta-amyloid or simply as amyloid, the protein was also seen to accumulate precipitously in the brains of people with Down syndrome, most of whom developed dementia by midlife. Glenner and Wong wondered if it was possible that an excess of this same protein set the dementia cascade in motion. They determined, after purifying the protein and sequencing its amino acids, that amyloid-beta was the product of a gene, and surmised that this gene was located on chromosome 21, the same chromosome affected in Down syndrome. Rival teams of scientists raced to find the matching gene. In early 1987 four independent research groups reported the discovery of a gene called APP, for amyloid precursor protein, on chromosome 21. This suggested strongly that they were on the right path. Investigators began looking to families with hereditary early-onset Alzheimer's disease—only a few of which were known at the time—to learn whether they carried abnormalities on the APP gene. If they did, it would validate the emerging hypothesis that amyloid was at the heart of the disease.

The late 1980s marked a pivotal moment in Alzheimer's research, offering the first snapshot of the proteins and genes involved and the first theory of how the illness developed, a theory that came to be known as the amyloid hypothesis. Twenty years later, Lopera would begin planning for a landmark test of the amyloid hypothesis in Colombia, but its significance largely missed him then. Having just escaped from a university under siege from right-wing paramilitaries, he was now treating children with learning disorders and autism. In the Brussels hospital where he worked, Lopera struggled with French and tried and failed to interest anyone in his genograms of Colombian farmers with dementia.

He enjoyed the peace of Brussels, where he gardened for his landlady—gardening would become his lifelong hobby and refuge—as a way of reducing his rent. In the summers the university arranged camping trips for its cash-strapped fellows to the United Kingdom and to Greece, where Lopera found himself misty-eyed at the sight of Mount Parnassus and the ancient ruins at Delphi, things he never imagined he would see, at least not at this stage of his life.

But he found the clinical experience in Belgium dull and circumscribed. Young doctors weren't given as much latitude as he was used to in Colombia, where you were thrown into the deep end early. Lopera was surprised to discover that he had already seen more patients with rare and difficult to diagnose diseases than many of the veteran physicians in Brussels. During grand rounds, meetings where perplexing cases were presented, he identified a case of palinopsia, a type of visual disturbance, in a boy who had been eating lunch at a restaurant with his mother when he saw the menu suspended in the air. "Everyone had different ideas, but I knew it was palinopsia because I'd encountered it at my hospital, for the huge number of patients I'd already seen," Lopera said. In Belgium he published the case, the first to describe palinopsia in a child.

Lopera dreaded the end of his fellowship and his return to Colombia. The Medellín cartel, in its escalating wars against its rivals and the Colombian government, had adopted terrorist tactics, kidnapping civilians and bombing hotels, military bases, banks, and power plants.

Colombians, accustomed to a hypersocial life spent outdoors in lively streets, retreated inside. In 1989, when Lopera left Belgium, he did not fly home immediately. Instead he flew to Miami and holed up with a Colombian colleague who had moved there. Three months later, Lopera finally landed home to face an eerie, tense city, where his colleagues no longer felt safe eating in restaurants. "Life became much more private," he said. "Everyone held their gatherings in their houses. We avoided going out for fear that a bomb could explode anywhere."

With trepidation he resumed his post at the San Vicente hospital. Lopera was unsure what would happen with the Alzheimer's study that had begun five years before with Pedro Julio Pulgarín. "It was neither my main interest nor my priority," he said. But if Belgium had taught him anything, it was that Colombia offered richer grounds for study than he'd previously understood, and that he was already a far more seasoned clinician than he had imagined himself to be. Lopera was not yet confident about his ability to collect and analyze data in a systematic manner. "What I was, was a curious doctor," he said. All kinds of cases streamed through the neurology department at San Vicente, and Lopera had eclectic interests. "We did a lot of cerebral trauma, cognitive rehabilitation, neuropsychology of learning disabilities, autism—we saw a lot of kids." Lopera published papers on phantom limbs, impairments of language and speech, facial recognition, and attention deficit hyperactivity disorder. He collaborated with Catholic theologians on a study of exorcisms, which were performed regularly still in Medellín, to learn which neurological diseases were prevalent among the ostensibly demon-possessed.

MANY PAISAS CONSIDERED 1989 to be a low-water mark for their city and country. The 1990s demonstrated that things could in fact get worse. Medellín grew only more violent, especially in the working-class neighborhoods on its periphery, places where country people who didn't want to pick coffee anymore, or who had been driven off their farms by guerillas, had been settling for generations, building wooden cottages balanced on piles and hoping for a better life for their children. Only now these children were being recruited by Pablo Escobar and his rivals as foot soldiers, cop killers, detonators of bombs. They murdered one another by the thousands in turf wars that would persist long after Escobar and the other big capos were dead. In the San Vicente hospital, its skilled emergency room staff received their bleeding bodies at a rate of ninety or so a night on weekends, saving as many as they could and leaving the rest for the medical examiner.

From downtown Medellín, where Lopera worked, you could see the neighborhoods spreading upward, turning the once-green mountainsides into erratic quilts of improvised housing. Every year the comunas inched higher until one day they would leave only the crests of the mountains untouched. It was from these places that the next wave of early Alzheimer's patients arrived.

Rodrigo was forty-three, one of nine children raised on a hillside south of Medellín that could have passed for a rural hamlet but was close enough to the city to afford Rodrigo access to something his montañero parents never had: a high school education. Rodrigo was working as an administrator in a large hotel when he fell into smoking *basuco*, a cheap, toxic byproduct of cocaine common in Medellín's streets. Though Rodrigo eventually broke his addiction, he never worked a real job again. He was reduced to selling lottery tickets until he had to stop working altogether, as he'd been giving customers the wrong change. Rodrigo didn't believe there was anything wrong with him except the occasional lapse in memory, or misplaced item. But his mother, who accompanied him to Lopera's clinic, insisted otherwise. Rodrigo was behaving just like her husband, who had died young with dementia. One of Rodrigo's aunts had also died with it. His grandfather, like Pedro Julio Pulgarín's mother, had died in Medellín's public mental hospital.

Rodrigo presented himself well and spoke well; his mini–mental state exam, a standard dementia screening test, was normal. But he had a hard time copying a Rey–Osterrieth Complex Figure, a bizarre-looking line drawing used since the 1940s to help diagnose dementia. Even a simple image of a daisy he couldn't reproduce very well. And yet Rodrigo had been a talented draftsman in high school, filling notebooks with careful studies of light, shadow, dimensions, and human musculature, alongside sophisticated notes on drawing theory. His family presented one of those notebooks to Lopera, who added it to Rodrigo's file.

This was the first time Lopera had an opportunity to observe a patient in the early stages of this heritable dementia, which he continued to suspect was a form of Alzheimer's. A CT scan revealed that Rodrigo's brain had begun atrophying. Over the next two years, Rodrigo deteriorated before Lopera's eyes. He drew a daisy with a human face, and on one visit he described his occupation as a pig slaughterer, something he'd done years before and only briefly. He began to suffer convulsions. At forty-five, he was diagnosed with dementia.

OFELIA, FORTY-THREE, HAD unmistakable symptoms of dementia when she came into San Vicente. She was bent over, disoriented, slow of speech, and walking strangely. Until recently a vivacious woman who attracted men half her age, she could no longer wear her high heels. Ofelia lived in a hillside comuna in the north of Medellín that had received thousands of families uprooted from rural Antioquia.

She had been a loving and capable single mother, a nondrinker and nonsmoker. When she was in her midthirties, working as a messenger, she would fall into crying fits without reason and announced plans to marry that she never mentioned again, as though she'd forgotten them. For many years she had suffered severe headaches.

Ofelia was fired from her messenger job because of all the errors she made, leaving her no way to support herself. Her sisters were alarmed to find her collecting bottle caps and other refuse from the street and bringing it into the house. She asked them for pen and paper with which to write a letter, but was able to produce only a sad, looping doodle.

Ofelia's father had been in a similar state—and just a few years older than Ofelia—when he was hit by a truck and killed, the sisters reported. Their grandfather had also died "crazy."

The neurology nurse Lucía Madrigal, who had worked with Lopera on the Pulgarín family, began building out Ofelia's and Rodrigo's genealogies, working by hand on paper taped together to form long scrolls. In most of the Spanish-speaking world, people use four names: a first and a middle name, then the father's surname, followed by the mother's. This helps trace links along the maternal as well as paternal lines. Madrigal was unable to discern whether the two patients were related, and they did not appear to be part of the Pulgarín clan in Belmira, whose genealogy was well established. But the three family trees revealed a small number of surnames common among them, suggesting a link in the past.

Ofelia's and Rodrigo's families had migrated to Medellín from the same region, hill towns close to where both Lopera and Madrigal had grown up. It dawned on Madrigal that when she was a girl, she'd heard schoolmates talking about people with problems that sounded like dementia, but that term was never used. Rather, victims were considered to have become stupid or crazy, cursed by malevolent priests, or *enyerbado*—under the influence of witchcraft. Madrigal interviewed as many of Ofelia's and Rodrigo's relatives as she could, collecting more reports of people who had started forgetting in their thirties and forties, and the researchers were astounded to realize how many cases there were, all with roots in the same handful of towns.

There was little consensus in the families as to what this was. They did not necessarily consider it a biological phenomenon or even a disease. They did not always distinguish its symptoms from madness. The explanations they had for it appeared to come from another century. One rumor held that you could get it by entering a certain cave. Or by reading the entire Bible, cover to cover. Or because you were lured by a flickering flame to a guaca, a cache of old Indian bones and gold. These were the sort of experiences that could stun you into this state forever.

THREE

The American scientist Ken Kosik crossed paths with Francisco Lopera in the early 1990s, just as Lopera was beginning to piece together the startling scope of his families with Alzheimer's.

Kosik had a slightly eccentric air about him; he wore his curly hair long and dressed as casually as he could get away with. A physician by training, his journey into medicine had been haphazard and reluctant. He'd grown up in a working-class district of Philadelphia, the grandson of Russian Jewish immigrants. His mother taught in a public high school; his father was a dentist "who hated his job. He rarely had patients and his office gathered dust." Kosik studied English literature in college and liked it enough to go on to earn a master's degree, but he felt he could do no more with it. He harvested vegetables, worked nights at a postal processing facility, and drove a taxi. Finally, "I decided to do something practical. But I did it in a very halfhearted way," he said. He applied to medical school.

Kosik found it hard to relate to the single-minded ardor of his fellow students at the Medical College of Pennsylvania. He took months off to hitchhike across Mexico, and devoted much of his time to performing with a theater group. In 1977, when soon-to-be graduates had to match with their residency programs, Kosik was among the handful of students not selected, but an opening soon became available at a small hospital in southwest Boston. "It was during that time, in the internal medicine department in direct contact with patients, that I became very serious and deeply involved in medicine," he said. Of all the specialties he explored, neurology interested him the most. He applied to different neurology programs, all of which turned him down, until an opening appeared at Tufts University, and Kosik was allowed in.

During a grand rounds lecture at Tufts, a neurologist named Dennis Selkoe presented a slide that captivated Kosik. The slide showed a gel electrophoresis of proteins in a brain from an Alzheimer's patient, compared with those in a healthy brain. Electrophoresis is a technique in which a sample of tissue is mashed up and dissolved with chemicals, then loaded onto a tray of gel. The gel is charged with an electric current, causing the proteins in the tissue to move through it and separate by molecular weight, forming bands. "And then you stain it with this brilliant blue dye. And it looks really beautiful. It's just exploding. And it was such a different way of looking at the brain," Kosik said. As a neurology resident, Kosik had already examined countless brain tissue slides, but the gels allowed for "a new level of analysis, something much more abstract."

Selkoe was working then at McLean Hospital, a sprawling psychiatric campus that got its start in the nineteenth century as the Asylum for the Insane. McLean was part of Harvard, which gave Selkoe access to autopsy brains from the Harvard hospitals and the wherewithal to apply what were then high-tech approaches to the problem of Alzheimer's disease. These included electrophoresis and immunohistochemistry, which used antibodies to identify disease proteins in lieu of traditional stains.

"We were fascinated by the idea: Could we get a molecular handle on how Alzheimer's began?" Selkoe told me. "Ken approached me right after the lecture, really interested in what I was talking about."

Kosik asked to join Selkoe's lab at McLean, but Selkoe couldn't hire him because he lacked the budget to carry a full salary. So Kosik sought a grant from Nancy Wexler, a young biologist already famous for her work in Huntington's disease, a heritable movement disorder that can also cause dementia. Wexler was working against the clock: her mother had died of Huntington's, a condition she stood to inherit, and she wanted to cultivate emerging investigators who might make their careers in the disease. Wexler "decided to take a chance on me, a guy who was a total risk with no laboratory experience," Kosik said. Installed at Selkoe's lab with Wexler's money, Kosik began running gels to analyze proteins in the brains of people with Huntington's and Alzheimer's.

By the mid-1980s, amyloid-beta had been identified as a core component of the plaques seen in Alzheimer's, but the tangles remained elusive. Researchers had long suspected them to be a corrupted form of one of the proteins that made up the structure of neurons. Kosik and Selkoe extracted tangles from Alzheimer's brains, resulting in a mysterious gunky substance that they injected into rabbits to see what kind of antibodies the rabbits produced. In the meantime, Kosik continued running his gels on the substance, searching for answers in the bands. He showed his results to some veteran cell biologists, who informed him that his gels showed a protein called tau. And to his and Selkoe's amazement, "the main antibody the rabbits produced was to tau."

Scientists had identified tau some years earlier as a component of healthy brain cells. It is one of the proteins that make up microtubules, which help give cells their structure and allow molecules to move around inside them. In Alzheimer's disease, Kosik and Selkoe proposed, tau becomes chemically altered and the microtubules collapse. Having unstable microtubules is like having damaged railroad tracks: nothing can reach its destination, and the neuron dies.

Kosik learned how to perform DNA sequencing and began sequencing tau, in a way similar to what other Alzheimer's researchers had done with amyloid-beta. "For a guy without a PhD, I really learned how to do some science and got immersed in so-called legitimate science," he said. When his results aligned with those of another scientist working with tau, "I was exhilarated," he said.

In the Alzheimer's field, though, it was amyloid that commanded greater interest. Selkoe, despite having made some early and important discoveries about tau, believed that amyloid had a more central, fundamental role in Alzheimer's, and therefore ought to be his lab's priority. "It was known as early as the fifties and sixties that tau tangles occurred in maybe a dozen brain diseases," Selkoe explained, "whereas amyloid plaques were more specific for Alzheimer's. I decided that amyloid deposits were more of an initiating point, and I still believe that today."

When Selkoe moved his laboratory from the McLean Hospital to the main Harvard Medical School campus, Kosik started his own lab, down the hall from Selkoe. Their parting was amicable, and they continued to collaborate for years afterward. But their differences as to how they viewed the principal proteins involved in Alzheimer's disease—amyloid and tau—would persist for the rest of their careers.

WHILE HE WAS at Harvard, Kosik became close friends with Enrique Osorio, a Colombian neurosurgeon who shared his interest in books and theater. As Osorio's fellowship wound down and he prepared to return to Colombia, he and Kosik decided to do "the kind of thing you only do when you're young—we said, let's write a grant to develop neuroscience in Latin America! How presumptuous!" But the National Institutes of Health awarded them some money, and in 1989, Kosik traveled to Bogotá to visit hospitals and researchers in the hope of establishing collaborations. Few American physicians would likely have made such a trip, given the rampant violence in Colombia then, but Kosik had a little bit of Spanish and a wanderlust that tended to supersede such concerns. He had been married since his twenties, but that marriage was beginning to falter, leaving him more primed than ever to get out of Boston.

Kosik took a room at the Hotel Tequendama, in a seedy, central part of Bogotá. The hotel faced a bullfighting ring and Monserrate, a pine-covered mountain that rises sternly over the city's eastern edge. The next day, Kosik and Osorio spoke at a neurology conference at the National University of Colombia, where Kosik became bored with the speeches and rounds of cocktails, eager to see more of Bogotá. From the little he'd seen so far, he found it darkly tantalizing: the student radicals trying to blockade the gates of the university, the riotous screams of the soccer fans as the Colombian team played Paraguay, the downtown building that had collapsed after a recent narco-bombing. Everywhere around were scruffy homeless youth whom people called *gamines*. Many were addicted to smoking basuco or inhaling solvents, and they slept in large makeshift encampments, where charity trucks showed up at night to feed them.

The elite neurologists hosting the conference discouraged Kosik from visiting Bogotá's public hospitals that served the poor, but Kosik was eager to. A Colombian resident led him into the bleak open neurology ward of the Hospital San Juan de Dios, with its iron beds and paint peeling from the bedstands. Kosik was taken aback by the absence of equipment, and the types of cases he encountered, which were almost inconceivable to an American physician. He met a twenty-year-old patient with a paralytic dementia from untreated syphilis, a thirty-six-year-old with cerebral toxoplasmosis, and a forty-year-old with a bacterial meningitis that had been left untreated for too long. Down a long hallway with arched windows and darkened rooms rested a row of women who had just delivered babies. Their infants were all lined up to begin life among the poor in Colombia—how many of them, Kosik wondered, might end up gamines?

Kosik also made a series of obligatory visits to the country's top scientists, who operated on a plane of prestige that was as alien to him as the decrepit halls of San Juan de Dios. He toured the sleek modern laboratory of Manuel Elkin Patarroyo, a forty-five-year-old immunologist celebrated for a synthetic vaccine for malaria that he had developed, testing it in Colombian soldiers. Patarroyo had published the successful results in *Nature*, but the World Health Organization was keeping cautious tabs on his vaccine, waiting for findings from bigger trials in other countries before calling it a breakthrough. In Colombia there was no such reserve. Patarroyo had so thoroughly impressed politicians that his work would come to claim more than half the country's medical research budget. "He was charming, like everyone in Enrique's circles," Kosik said. "I was sort of impressed. I didn't know he did all kinds of evil shit." A few years later, after bigger trials failed in Thailand and Africa, Patarroyo's malaria vaccine was found to be no better than placebo.

Kosik also called on Jorge Reynolds, an engineer celebrated throughout Colombia for having invented the cardiac pacemaker in the 1950s. Reynolds, aging and independently wealthy, maintained his offices in his enormous home, where Kosik passed uniformed women and security men in suits before reaching Reynolds's door. There he encountered something akin to a museum. A cabinet contained an array of old pacemakers, and glass-enclosed shelves displayed skulls from Peru upon which Inca surgeons had operated. In the center of the floor stood the first pacemaker Reynolds had designed, a weighty contraption balanced on a dolly and attached by wires to a car battery.

Reynolds, like Patarroyo, would also one day be exposed as a liar. His contribution to the invention of the pacemaker turned out to have been marginal. But in that moment he was considered a national treasure, and he welcomed Kosik warmly. In his strange office-museum they talked about Reynolds's expeditions to get cardiograms on whales, for which the Colombian navy had lent him submarines. Reynolds and Kosik talked at length about brain anatomy, an interest they shared, and Reynolds presented Kosik with a special gift to take back with him to Harvard: the brain of a pink dolphin collected in the Amazon.

MEDELLÍN'S INTERCONTINENTAL HOTEL is a cold, citadel-like structure perched on a hill. Its high level of security, including men with radios everywhere and secret entry codes known only to the trusted cabbies allowed through its gates, is conspicuous even today. In late 1992, when Ken Kosik checked into his room there, lay on a bed, and stared at the wall, the city's most notorious resident, Pablo Escobar, had escaped from prison and launched his final, ill-fated battle against the Colombian state. A year later Escobar would be lying dead of bullet wounds on the terra-cotta tiles of a rooftop in Medellín. For now, though, he was at large, his whereabouts unknown, and car bombs were going off.

Kosik was exhausted, at a breaking point. He recently had moved out of his family home in Boston and into a place of his own, where the few belongings he'd taken with him lay strewn all over. Only two nights before, he had been shocked on returning from his lab to find a policeman in his living room, investigating his estranged wife's false allegations of abuse. At this point there was no possible reconciliation; seeing his children required negotiation and his former house was off limits. He felt safe in Medellín, a fact whose irony was not lost on him as he sprawled out on the hotel bed.

Kosik had flown to Medellín with his colleagues from Bogotá for a neurology conference, which was being held that weekend in the same hotel. What he didn't realize about his brief trip to Medellín was that it had all been hastily arranged, down to the little neurology conference, so that he could meet Francisco Lopera. A group of brain researchers in Bogotá, friends of Enrique Osorio, had just days earlier hatched the plan.

"Medellín was in a very complicated state. We knew the city had a strong stigma, so we decided to propose the trip to Ken only once he got to Bogotá," Diana Matallana, the neuropsychologist who led the effort, told me. "We'd be super cautious. We'd go straight to the conference hotel and twenty-four hours later be back." A top lieutenant of Pablo Escobar had just been killed by the police and the city was militarizing in anticipation of a vicious response. On the plane to Medellín, Kosik's colleagues removed all newspapers from his sight.

The Colombian scientists were isolated and they knew it. In an age before the internet, with costly journal subscriptions and international conferences still the only way to know what was going on in medicine, a visit from a Harvard Alzheimer's researcher was too precious an opportunity to let slip. At the conference Kosik lectured in Spanish about Alzheimer's disease, or tried to, reading his text from a piece of paper. "And Lopera was in the audience," he said, "and afterward he came up to me and he said, 'Doctor, I have this large family that seems to be getting dementia.' At the time we didn't know if it was Alzheimer's, or if it was even genetic, though it looked like it had some inherited pattern. And that changed my life, of course."

FOUR

Lopera was slightly behind the curve when it came to what other research teams were doing. But he knew he was onto something of potential importance with the paisa families. Alzheimer's research had taken on new urgency worldwide. Part of that was because the concept of aging itself had changed, especially in wealthier countries, where retirement had become synonymous with enjoyment and leisure. Dementia ruined all that. The perception of dementia as an inherent possibility of aging had given way to that of a dreaded disease—a "major killer," wrote the researcher Robert Katzman in 1979—worth battling at all costs.

In the United States, high-profile cases like that of the actress Rita Hayworth helped the nascent Alzheimer's Association raise awareness and funds. In 1983, the U.S. House of Representatives declared November to be the first Alzheimer's Disease Awareness Month, and the following year the National Institute on Aging began building a network of research centers dedicated to Alzheimer's. The convergence of new research methods and heavy funding created a boom in interest that drew talent from many branches of science: genetics, molecular biology, pathology, psychiatry. And all over the world a race was on to study more early-onset Alzheimer's families. The hope was that studying their unique genetic mutations, and learning how these acted to trigger the disease, would pave the way to therapies or even cures. The amyloid precursor protein gene, reported in 1987, was only the first to be identified. Researchers now understood that this wasn't the only Alzheimer's gene or even the most important, because new families had been found that had no mutations on APP.

Lopera aimed to deepen his work with the Alzheimer's families, and the University of Antioquia backed his efforts. A silver lining of Colombia's drug and political violence was the birth of a new national constitution, in 1991, that set priorities for the government, some of them downright idealistic. Promoting scientific investigation was one. At the university, where churning out doctors had been the focus for as long as anyone could remember, administrators were putting together new multidisciplinary research teams. Jorge Ossa, the biologist leading them, considered Lopera's work in Alzheimer's disease to be among the most promising investigations the university had going, and sought to build a broad team around it.

As fellow researchers from rural families, Ossa and Lopera had a strong affinity. Ossa's parents grew coffee, and seven of his eight siblings never made it through high school. Like Lopera, he had been slated early on for the priesthood, a common outlet for young rural men with intellectual tendencies. "If I'd told my family that I wanted to be a scientist, they would have thought I was nuts," he told me. Ossa spoke English and could act as a bridge to outside interests. But he didn't know what to make of Ken Kosik when he showed up in Medellín. Whatever this Alzheimer's study was to become, "we needed this to be a Colombian project, and not something stolen by the gringos," Ossa said.

Lopera, however, took to Kosik instantly. In Medellín the two physicians, limited by language, scribbled graphs on papers and borrowed bilingual colleagues to help them put their ideas together. Kosik struck Lopera as a rare type: "a cool neurologist," maybe something of a hippie. Lopera displayed a humility that Kosik realized was often lacking among Colombian scientists. Rather than hoard his data until he could announce some earth-shaking discovery, in the style of Manuel Elkin Patarroyo, Lopera was anxious to have the outside world scrutinize his work.

Kosik confirmed to Lopera what Lopera had long suspected—that there likely was an Alzheimer's gene to find in Colombia. Kosik told Lopera he had the tools to at least begin looking. In the meantime, they needed more blood. They needed to organize and standardize genealogies that were growing unruly. Most of all, they needed a brain from someone who had died, to prove once and for all that the Colombian families' condition was in fact Alzheimer's disease and not another form of dementia.

Ossa secured the group its first grant, for a scant eight hundred dollars. All they really required at that point was money for food and gas. Instead of traveling to the hills of Belmira, where the Pulgaríns tended their cows, the investigators focused on a place where the dementia cases were even more abundant: a rural hamlet called Canoas, which lies on the road from the large town of Yarumal—where Lopera and his colleague Lucía Madrigal had both grown up—to the smaller town of Angostura, twelve miles away. Many of the young dementia patients who had come to the hospital in recent years seemed to come from this spot, or could trace their origins back to it. In Belmira, Lopera and his colleagues counted eleven dementia cases, in Yarumal another eleven. In tiny Canoas, they counted fifty-two.

THE HAMLET OF CANOAS got its name—which means conduits—in colonial times, when settlers chopped down the silver-leafed yarumo trees that grew there and hollowed out their trunks to shunt river water to their homes. These days it is a restful place with porches well swept, hanging plants tended, and green coffee beans spread out to dry in the sun. Even the poorest houses are painted in bright contrasting colors, touched up every year. When the guavas growing on trees along the roadside are ripe, the air fills with their scent, and cows tug the branches down with their teeth to get at the yellow fruits. Red *chivas*, old Ford trucks retrofitted as buses that carry people and goods through the mountains, stop at Canoas's only tavern, where a small statue of Padre Marianito, in his black cloak and hat, greets visitors on that same curve of road. It was here in Canoas that Marianito, a local priest revered as a saint long before the Vatican was persuaded to beatify him, performed his best-known miracle. There are different versions of the story, but it is generally said that Marianito was on his way to see someone sick when he encountered a man who had lost his faith. The man pointed to a dried-up guava tree and told him that he would return to the fold only when the dead tree sprang back to life. Weeks later it did, and the man came to Marianito breathless with emotion, his faith restored.

Francisco Piedrahita was an adolescent in Canoas when Lucía Madrigal started coming around with her questions and test tubes. When I met him, he was in his late thirties, a neurology nurse who worked closely with Lopera and Madrigal. With his frenetic, deeply empathetic manner, Piedrahita was adored by his colleagues at Neurociencias but most of all by the Alzheimer's families, who recognized him as one of their own. His grandparents, aunts, and uncles had died of early-onset Alzheimer's before anyone knew what it was.

Piedrahita's dual role—as an investigator and family member—made him invaluable to the scientists, a bridge between two worlds. Piedrahita prayed to Padre Marianito when he needed a favor, just as the families did. He knew what it was to become so isolated caring for a person with dementia that your own mental health suffered; caretakers often became depressed, anxious, and afraid to leave the house.

The family members trusted Piedrahita like they trusted no one else. They let him draw fluid from their spines and distract them with cheerful patter while they were injected with radioactive chemicals to light up the amyloid in their brains. Piedrahita escorted groups of them periodically to Boston, where colleagues of Lopera's conducted more advanced experimental imaging. He turned each trip into a party, an opportunity for pizza and shopping, never talking about Alzheimer's.

We sat in a consulting room one afternoon as Piedrahita told me about his childhood in Canoas, his rapid speech reverberating off the bare walls.

In the late 1980s, he told me, the road to Canoas was unpaved and treacherous and the chivas passed infrequently. In his parents' generation, Piedrahita said, "you left the village three times: the day you were baptized, the day of your first communion, and the day you married." It went without saying that whomever you married was, to a greater or lesser degree, a relation. Piedrahita's parents were first cousins, leaving him and his siblings with the double surname strikingly common in the region: Piedrahita-Piedrahita. For the first six years of his life, the finca lacked electricity, and the only telephone in the hamlet was a communal one, where the family waited on Saturdays, excitedly, for a call from an aunt in Medellín. Right-wing paramilitaries had yet to become a force in Canoas, and the FARC guerillas, who lurked in the surrounding mountains, were no more than phantoms to a child. "All we did was play," Piedrahita said. "We played with anything we could find. We tied ropes to the trees and made swings. It was a delicious life, a very different kind of life, and we didn't know anything else."

When Piedrahita was five years old, his maternal grandfather wouldn't stop carrying him around. "Put the boy down," everyone told him, only to watch him do it again, as though he hadn't heard or couldn't understand. His grandfather was forty-seven at the time.

Three years later he was confined to a bed. When his mother sent Francisco on errands to the neighboring farms, he often came across someone disoriented or bedridden. Once he entered a house and found himself alone with a man tethered to some furniture by a rope around his waist. The experience shocked Piedrahita, and he asked his mother what was happening. "It's an illness that people get," she told him, "and you'll come to understand it one day."

Piedrahita's mother, Ledy, was something of a freethinker for Canoas. She wanted her children to be educated even if it meant having to send them to live with relatives. Her opinion on the local affliction—that this was a medical condition—was a minority one. The conventional wisdom in Canoas held that it was witchcraft, which could mean a lot of things. If the sick person was a man, it might mean he had been cursed by a jealous woman. If it was a woman, especially a sexually adventurous one, she might have been cursed by a priest—priests were alleged to inflict the cruelest kind of curses, ones that lasted for generations. Witchcraft could also mean poisoning. "If you had at any time in your life a problem with your neighbor, it was your neighbor. She worked on you, she gave you some god-awful thing to drink. Or if a man wanted a woman and couldn't have her, then he'd make sure nobody could," Piedrahita said.

It wasn't such a stretch to think that someone in a disoriented, nonsensical state was enyerbado, under a spell; herbs made up the essential pharmacopoeia of the countryside since the time of the Conquest, when European formulas used since the Middle Ages met with those of Indigenous Americans and Africans, in an explosion of the healing arts and of magic. Everyone knew that with the right herbs you could abort a fetus, purge a bad stomach, calm nerves, and quell pain; you could just as easily poison a lover or keep one close. Much of the witchcraft practiced in Antioquia built on herbalism: spells and incantations, which local people simply called "praying," were used to give power to herbs, to propel their properties into the realm of the supernatural. "No one ever sought a doctor," Piedrahita said. If they brought their sick anywhere it was to "the lady who said prayers." Everyone knew someone who prayed for a fee, and in much of Colombia today, everyone still does.

Piedrahita's paternal grandfather, whom he never knew, left the family to work, and ended up living with another woman. Years later they learned he had "gone crazy" and been sent away. People with dementia were as frequently described as crazy or *bobo*, "stupid"; a long period of inexplicable behavior often preceded the forgetting. Piedrahita's grandfather had died, it turned out, in Medellín's mental hospital. No one could locate his remains to return them to Canoas, and he was buried in a potter's field in a city he never knew.

By the time Piedrahita left Canoas, at age eighteen, his grandmother was showing the first signs of dementia; she was putting sugar in the soup, he said, mistaking it for salt. Because families were so big, generations often overlapped, and people had aunts and uncles their own age. One of Piedrahita's aunts was dying of Alzheimer's around the time his grandmother fell ill. All of his father's siblings, and many of his mother's, would go on to develop the disease.

Piedrahita borrowed a sheet from my notebook and scrawled out a genogram: a confusion of black squares and circles, death and sickness in all directions. By some stroke of luck, it appeared, both of his parents had been spared.

I told Piedrahita that it seemed he and his siblings would not get sick, that despite the bad odds, the chain had been broken. Piedrahita said he hoped so, but that there were rare people in the families who got sick late—sometimes more than a decade later than they were supposed to. Perhaps one of his parents would, too.

He crumpled up the genogram, tossed it in the wastebasket, and continued his story.

When Piedrahita first met Lucía Madrigal in Canoas, she was "young and very beautiful," he remembered, carrying a bunch of tubes to take blood samples. "She told us there were a lot of people here with an illness—and that this was an illness—and that they were going to see what people had and what it was called."

Though Madrigal had grown up in Yarumal, just a few miles away, she was a stranger in Canoas, where guerillas lurked out of sight but maintained their networks of informants. Looking back, Piedrahita said, it was a miracle that nothing terrible had happened to her. Madrigal came on weekends, taking "blood, blood, and more blood," Piedrahita said. Once, she took blood from Piedrahita's father after she came across him leading mules in a ravine. People started rumors of what she was doing with all the blood.

Piedrahita's mother was not among the skeptics. Piedrahita remembered her holding his brother, then only a toddler, so that Madrigal could draw blood from the boy. For Ledy Piedrahita, the revelation—coming directly from a nurse at the University of Antioquia—that they were dealing with a neurological disease and not some epidemic of spells was immensely gratifying. "I told you it wasn't a curse," she chided Piedrahita's father, a peasant cut of a more traditional cloth.

Madrigal had been studying psychology, a useful background when going from house to house in a closed and superstitious region, posing intimate questions and asking for blood. Constructing a genealogy was a painstaking, delicate process that could reveal secrets that most people would rather have taken to their graves: relatives who died in bizarre and humiliating circumstances, or were abandoned when they became ill; children who were born of incest, rape, and affairs.

Madrigal learned that there were ways to elicit the necessary information without offending. You learned to sit next to the person, not across from them, as you sketched the tree. You interviewed family members separately and in private, repeatedly if you had to. You always wanted to start with the oldest person you could find. Madrigal was careful to avoid the worst mistake you could make: putting questions to someone who was sick, even just barely. More often than not, sick people did not recognize that they were sick.

She began to draft a family tree of the Piedrahita clan, the widest and deepest of all the genealogies under way. It comprised then a bunch of handwritten pages taped together into a long scroll—"a horrible roll of paper," Piedrahita said, because it showed in no uncertain terms the scope of the disease, the steady and merciless march of it over time. But Francisco Piedrahita was also moved by that long document, which Madrigal sometimes showed journalists or displayed at the research group's events, because it represented the sacrifice of Madrigal's youth, all her efforts in the countryside.

Piedrahita and Madrigal were very close, though their personalities couldn't have differed more. Piedrahita was gentle, Madrigal queenly and tough. Between the two nurses they retained a cohort of thousands—first names, last names, faces—in their heads. They knew what villages everyone's grandmother came from, what barrio of Medellín they lived in and under what conditions. They traveled to the homes of the sickest patients, offering aid and politely prodding caregivers to donate brains when the end came. This kind of research required a balance of the compassionate and the transactional that they, and Lopera, had honed to perfection.

Piedrahita believed deeply in what the research group did; his stake in its success was personal. "If Neurociencias hadn't existed, this would be a forgotten disease," he told me. "We didn't know what was happening." Lopera and Madrigal "were the only ones to go through the trouble to come into the hamlets where you have to get a chiva, you have to take unpaved roads, where conditions are primitive. To put on rubber boots, to take a road you don't know to a place you've never been."

I mentioned to him what Lopera once told me—that a *médico rural* had been stationed back then in Angostura, the town closest to Canoas, but he had failed to notice anything amiss in the hamlets. Lopera had relayed this with dismay and amusement, but Piedrahita said he was not surprised. People went to the doctor only when they had a machete wound to be sewn up or their appendix was going to burst. When they had unbearable pain.

"And Alzheimer's doesn't hurt," he said.

AS THE GENEALOGIES SWELLED, Lopera became increasingly convinced that this was a time bomb, an epidemic in the making. These rural families with dementia were huge, with up to twenty children. They comprised more than a thousand people, many younger than forty. And there was no telling how many more branches were yet to be identified in remote towns whose médico rural had been asleep at the wheel.

Lopera turned to a geneticist colleague at the university to try to parse the magnitude of the problem. Mauricio Arcos-Burgos was enchanted by what Madrigal and Lopera were doing, and he knew about building family trees. Arcos-Burgos traveled with the team to the hill towns. The affected clans had a distinct and very traditional way of life, Arcos-Burgos observed, which suggested to him that they might be genetically close as well. The way the same surnames kept cropping up in the genealogies, Arcos-Burgos felt he was dealing with an isolated population with a small number of common ancestors—and a disease that had been present for a very long time.

He assigned each of the affected clans the letter "C," for Colombia, and a number. C1 represented the group in Belmira, C2 was Ofelia's clan in Canoas, and C3 was the Yarumal family of Rodrigo, the man whose once-formidable drawing ability disintegrated as his disease progressed. Using a clumsy, antiquated program on the university computer, Arcos-Burgos spent two months running models. The models confirmed that the causative gene, whatever it was, got inherited in an autosomal dominant pattern, passing from a parent to half of his or her children, as the genealogies already indicated. They also showed that this unknown gene, while still powerful enough to cause disease in everyone who carried it, did not determine everything about how the disease behaved. People with the gene would get the same symptoms, at roughly the same age. But there remained room for other factors, including other genes, to influence its timing and course.

In Boston, Ken Kosik began working with blood samples that Lopera had stored for years. Gel electrophoresis could be used to detect genetic mutations; you compared the same sequence of DNA from a healthy person with one from a sick person and looked for differences in the bands. But you had to know which gene to target. Scientists assumed by now that there were several genes involved in Alzheimer's disease, but so far only one—the gene called APP, on chromosome 21—had been linked definitively to its early-onset forms. It was on APP that Kosik launched his search for the Colombian Alzheimer's mutation. "If you know what the gene is, and you have blood from a patient with the disease, it's easy," he said.

Kosik's search turned up empty. This meant that a gene besides APP was in play. Researchers working with an early-onset family in Italy, which was also negative for mutations on APP, thought there might be another Alzheimer's gene on chromosome 14. Perhaps it was the same one affected in the Colombians. But Kosik had reached the limit of what he could do in his own lab. Launching a study to identify unknown genes "was over my head at the time," he said.

In 1994, Kosik, Lopera, Madrigal, and Arcos-Burgos published a paper that brought together what each had learned to that point about the disease. Without a single autopsy brain to confirm the diagnosis, they still couldn't call it Alzheimer's. Lopera and Madrigal presented the clinical findings they had accumulated for more than a decade. By then they'd seen and studied eighty-six young patients with dementia. The average age at which people developed dementia was forty-seven—just as Lopera had reported in his first study.

Several months after the paper was published, a woman named Florelba died, at age fifty-seven, in the hill town of Angostura. Florelba had been gravely ill with dementia when Lopera and Madrigal first evaluated her. Her family had agreed, at least tentatively, to donate her brain when she died.

The job of collecting it fell to Juan Carlos Arango, an overworked pathologist in the San Vicente hospital, out of whose wards the flow of gangster bodies never ceased in those days. Florelba's body lay in Angostura, more than four hours away. All afternoon, Arango said, he tried to plead with doctors there to extract the brain—something any pathologist knew how to do—and preserve it for him in formalin. Brains were more resistant than other organs to decomposition, and Arango estimated that he had a day or so to get this one out. But as evening approached, Arango realized that no one in Angostura was going to extract the brain for him. He'd have no choice but to go there himself. Arango, Lopera, and Madrigal set off in Arango's small car. In Yarumal they borrowed another vehicle, a 4x4, from the local hospital, arriving in Angostura at 1:00 a.m. as a thunderstorm roared around them. The group took shelter at the home of Florelba's family, where her grieving husband and children were gathered. But the team was not allowed to extract her brain right away: the siblings wanted to wait until one of their brothers arrived.

"We were in the years right after Escobar," Arango recalled, when Medellín's underworld was breaking into smaller mafias. As the morning wore on, "this guy arrives who looks like he's working with them." The brother was "the very image," he said: an ex-cop, gold chains and rings, sunglasses. Before he showed up, the family had been nice. "And then everything changed."

Florelba's children—Arango couldn't recall if there were six or eight—vacillated on their decision as they drank and wept around their mother's body, and the ex-cop son's bearing spooked Arango and Lopera so much that they retreated to Angostura's small hospital to regroup. Finally, with little left to lose, Arango called the house. "I told them, a little angrily, that I was leaving at noon. If they wanted me to do the autopsy, fine, and if not, that was it, no problem. I went out to get coffee—we were exhausted as we hadn't slept all night—and one of the daughters arrived ten minutes before noon and told us they'd said yes," he said. "She signed the papers. It was over. We sent an ambulance to pick up the body and deliver it to the hospital morgue." But when Arango and Lopera entered the morgue, they were shocked to see another of the sons waiting for them there. He had come to demand money, "saying we were going to sell the brain to the gringos," Arango said.

At that point Arango exploded. "I said something really indecent," he remembered, but it worked; the son left. Arango and Lopera opened Florelba's skull from behind, leaving her face and forehead intact, and extracted the brain, which they placed in a bucket with formalin. The brain had the overall atrophied appearance of Alzheimer's disease, Arango thought, but he wasn't ready to cut into it and find out; Lopera and Kosik wanted him to carry it whole to Kosik's lab in Boston. Florelba's brain was the only thing standing between the group and confirmation that the disease they'd been studying for more than ten years was, indeed, Alzheimer's.

It wouldn't do to take the brain in a bucket. "I went to one of those camping stores and bought a metal thermos," Arango recalled. He flew first to Miami, thermos in hand, and declared the brain to customs, as he figured he should. At the time, Colombian travelers were harassed by customs as a rule—their bags of coffee sliced open, their toothpaste tubes emptied. But once he'd said enough about Alzheimer's disease and Harvard, Arango and his brain were waved through.

In Miami he was received by some of Lopera's old friends and colleagues. One of them, an anatomist, wanted to borrow the brain for a study of his own before forwarding it to Kosik. Arango spent the whole morning listening as the anatomist tried to sell Arango on the idea. The brain was too important to leave behind, Arango felt, even just a chunk of it, as the anatomist and his colleagues proposed. "They had me practically locked in that office—they wouldn't even let me eat lunch," he said. Arango's flight to Boston was that evening and he insisted on leaving with the brain intact, aggravating his hosts. Arango did not reach Logan Airport until nearly midnight, where Kosik awaited him in an empty arrivals hall.

The next morning, on the Harvard campus, Kosik and Arango and a cadre of students and technicians and pathologists gathered around the brain, ready to make the first cuts. "The cuts were beautiful," Arango remembered. They preserved the sectioned tissue in paraffin, and treated thin slices of it with silver-staining methods not fundamentally different from those Alois Alzheimer had used some ninety years earlier.

The final report left little room for doubt:

> The most striking feature is a diffuse destruction of the cortex. Plaques are distributed throughout the cortex and hippocampus. There are prominent neurofibrillary tangles and "ghost" cells. The cerebellum has abundant plaques detected readily.

It was Alzheimer's disease that had destroyed Florelba's brain.

FIVE

The paisa genealogies looked to Kosik like collages: accordion printouts covered with handwritten corrections and additions, with church records and random scraps of information tacked all over them. Some were so long they had to be suspended by two people to read. Sitting around a table with Lucía Madrigal and Francisco Lopera in the San Vicente hospital in early 1995, Kosik noticed that certain surnames—Piedrahita, Pulgarín, Pineda—kept recurring and recombining. The first names struck him as having a musical quality, especially the women's: Gilma Respa, Alta Gracia, Rosa Fabiola. Long chains of circles and squares below the names showed that many of these women had given birth to a child every year over the course of her reproductive life, though a significant portion of those children had not survived.

In *One Hundred Years of Solitude*, Gabriel García Márquez had captured something essential about families over time, Kosik thought. The seven generations of the fictional Buendía family, living in the riverside town founded by their ancestors, are unable to escape their own history, the patterns set in the past. Looking at the genealogies of these Alzheimer's clans, the stories ceased to be personal. That level of resolution was lost. The family was the unit, Kosik thought, its branches growing and shrinking endlessly like fractals.

Life had improved for Kosik since his Bogotá days. His divorce was still in process, but he had a new partner, a Korean-born biologist, and they were expecting a baby. The Alzheimer's project in Colombia was evolving in a way that had become enormously exciting for him, and his collaborators in Medellín, Lopera and Jorge Ossa, were more down-to-earth than the physicians he'd known in the capital.

His mission in Medellín was urgent: *Get the pedigree, its interconnections, figure out what's going on, and bring back the samples*, he wrote in his diary. The team had the results from its first brain and a fair shot at identifying an unknown gene implicated in familial Alzheimer's disease. This was not something they had the resources to do alone, however, so Kosik had to recruit a geneticist capable of mounting the kind of costly, labor-intensive study required.

Geneticists had long understood that different versions of a gene, or alleles, located close to one another on the same chromosome will often be passed on together to offspring. In a large extended family with an inherited disease, they could track the inheritance of alleles whose locations were known, to see how frequently these co-occurred with the disease. This allowed them to narrow down the likely location of the gene. What was needed for this type of study, called a linkage study, were very large families.

Across the border from Colombia in Venezuela, Nancy Wexler—the biologist who had helped Kosik at Harvard—had been conducting a genetic linkage study in poor fishing families with Huntington's disease for more than a decade, constructing genealogies and collecting thousands of samples of blood, skin, and semen in her search for the affected gene. In the early 1980s, Wexler's team had narrowed down its location on chromosome 4, and within a year Wexler would announce the discovery of the Huntington's disease gene, making news all over the world.

Kosik felt that something similar was possible in Colombia. To find a major new disease-causing gene in that era was "a big prize," Kosik said, and it thrilled him and Lopera to think that they could. Kosik approached Alison Goate, an English geneticist who had made a splash a few years before with her breakthrough discovery of an Alzheimer's-causing variant on APP, which came to be called the London mutation. Goate had since moved to Washington University in St. Louis, Missouri, where her lab was working with a handful of early Alzheimer's families suspected to have mutations on a gene on chromosome 14. Goate had a reputation for openness in a field defined by intense rivalries. "There were a lot of different geneticists I could have invited to help us," Kosik recalled, "but they were all competing at the time and they didn't talk to each other. I felt Alison was the only one I could really work with."

Two genes were now confirmed to be involved in Alzheimer's disease: APP, which was implicated in early-onset cases, and APOE, which had since been found on chromosome 19 and was linked to late-onset cases. Which variants of APOE a person had, and how many copies, increased or reduced the likelihood of developing the disease as they aged. The third, unknown gene, which like APP caused early-onset Alzheimer's, was the one Goate was searching for on chromosome 14.

Goate was aware that other Alzheimer's researchers were hunting for the same gene. One group was seeking it in an Italian family, and while this was the biggest early Alzheimer's kindred described to date, it was still much smaller than Lopera's paisa clans. "If you assumed that the different families that we had from these different villages were actually the same family, you already knew it would be way bigger than any other," Goate told me. "I can't think of another where people had so many children."

In a linkage study, family relationships cannot be assumed based on genealogies, which were largely self-reported. People sometimes misrepresented the truth about their parentage, or didn't know it. You needed blood samples to show relatedness, Goate explained, and "if you have samples from eight, nine, or ten children, there's a good chance you can reconstruct the chromosomes in the affected parent who died."

Lopera and Madrigal had coaxed hundreds of blood samples from the families, but they'd been stored in refrigerators that had thawed and frozen over during power outages; some contained only dried-up residue. Goate sent Lopera money to get teams back into the field, collecting fresh blood. Lopera remembered the months that followed as "feverish." As the results were analyzed, the genealogies snowballed, revealing clusters of early Alzheimer's in towns not previously on the investigators' radar. When the families' information was first entered into a program for tracking inherited diseases, it crashed: the genetics software of the time was no match for Colombian Catholic megaclans. Endless genograms came to cover the conference room walls of Goate's offices in St. Louis. Goate emailed Lopera at all hours of the night demanding more details, more clinical histories, more blood.

As the gene hunt progressed Kosik traveled frequently to Colombia, where Jorge Ossa still viewed him with caution, unsure whether he and Goate were likely to steal credit if the investigation revealed a gene. But Lopera and Kosik were becoming ever closer friends. Lopera was upbeat, smiling, with all the trappings of a successful physician anywhere in the world: a charming wife and baby daughter, a spacious new condominium, a well-equipped office where he saw private patients once a week. He was agile minded and broadly curious, able to talk with equal ease about speech disorders and Noam Chomsky. He was the kind of neurologist who reminded Kosik why he'd gone into the field.

It was all the more impressive to Kosik that Lopera had carved out such a life in Medellín. While murder rates had dropped from their Escobar-era peak, the city still saw more than 4,500 killings a year. Lopera and his group worked out of a hospital system whose daily bread was carnage, where assassins would break into recovery rooms to put more bullets into an enemy who'd just been painstakingly sewn up. Lopera maneuvered adeptly within this world, Kosik saw. He traveled in his own car and avoided public spaces in a way that seemed perfectly natural. When Kosik was home in Boston, everyone wanted to hear his opinion on "the violence in Colombia." But here in Colombia that violence became background, part of a complex tapestry of life.

Lopera invited the Alzheimer's families living in Medellín to convene on the San Vicente campus so that Kosik could meet them, explain their gene-hunting effort, and ask them for blood. The room was already full when the doctors arrived, with people crowding the halls, healthy and sick alike. These were the urban Pinedas, Pulgaríns, and Piedrahitas, in their jeans and tank tops, and the whole affair had the buzzy feel of a reunion; many of these cousins hadn't seen one another in years. Among them were people at that crucial tipping point in their midforties, hovering between well and sick, their disease still undiagnosed but their families starting to sense that something was wrong. The sickest sat in wheelchairs, rigid and not speaking, with eyes roving and hands clenched. And yet their relatives had managed to get them here, down the steep hillsides of the comunas where so many of them lived, along with the children and teens who tenderly held their hands and kept them company.

The doctors each spoke briefly about Alzheimer's disease, explaining that the inheritance the families shared had a name, a known set of symptoms, and a defined progression. The families responded with questions. What were the investigators doing with all the blood? they wanted to know. Kosik wasn't accustomed to thinking of genes in everyday terms. Any discussion of them, for him, started from the fact that genes were made of DNA, which in turn was made of nucleotides, of introns and exons, whose structures and sequences encoded proteins that went on to regulate other proteins and the behavior of cells, in a complex cascade ultimately resulting in a trait. But he'd recently been asked to address his son's sixth-grade class about genes, and in doing so he settled on a simple, direct formulation: a gene made a trait. Kosik repeated that to the families: a gene made a trait, and in this case, the trait was Alzheimer's disease. They needed blood to find how the disease appeared, he told them, and to discover how, one day, they might find a way to stop it.

Lucía Madrigal and her assistants set up a table to take blood, and the families lined up to give it. As the doctors circulated from group to family group, Kosik was struck by Lopera's way with them: he was jocular, natural, and spoke their language. He saw something old-fashioned in Lopera, like a country doctor who still believed in the laying on of hands. This was a dying concept among physicians, but its value here was clear. Kosik and Lopera were promising no treatment, no cure; they were only asking for help. But the families seemed pleased to share a moment with someone who could appreciate their troubles.

Kosik, Lopera, and Ossa headed to Yarumal, to repeat the meeting with the rural branches of the clan. Madrigal had preceded them by a day, to get the word out and give the families from distant hamlets time to come in, by horse or mule if they had to. The team drove past the small church where Lopera's parents had wed, until the whole town appeared above them, nestled idyllically into the mountainside. But Yarumal only looked quaint. Guerillas and paramilitaries were both active in the region, meaning that the investigators could not tour Canoas or the other hamlets where the families lived, but instead had to ask people to make their way to the town hospital. Yarumal's mayor assigned the doctors a pair of bodyguards.

The group awaiting them in the hospital's auditorium was less animated than the one in Medellín. Village elders sat quietly in wide-brimmed straw hats, while the visibly ill were draped in wool *ruanas*, or ponchos, attended by their spouses and children. People referred to the disease as "the condition," and though Kosik did his best to explain autosomal dominant inheritance, many seemed perplexed as to why some families had more sick members than others. There were families with nine sick and three well siblings; there were families with ten well and one sick. Not everyone understood the mathematical concept of probability, Kosik realized. He explained that it was the same as having all boys or all girls. You had an equal shot at either every single time, but it could come out any which way.

The doctors were intrigued when a woman in the audience blamed a tree for the disease. "Physical contact with a tree," Ossa remembered. Ossa pressed the woman for more information on the tree: What did it look like? Where did it occur? But she could not say. "She told us, 'If I knew which tree it was, I'd yank them all up one by one,'" he recalled, leading him to think that the tree was something mythical, a symbol of her pain.

Meanwhile, in St. Louis, Alison Goate and her colleagues were scouring through the blood samples Kosik and others had flown up from Colombia. They looked for variants or mutations—changes in the DNA sequence—in the region of chromosome 14 where the Alzheimer's-causing gene was thought to occur. And then they looked to see whether the sick individuals shared the same haplotype—a chunk of DNA that passes from generation to generation as a unit—around the spot. The researchers found that the haplotype corresponded with the variant, every time. This meant that the Pulgaríns,

Pinedas, Piedrahitas, and all the rest were one family. They had inherited their disease from a common ancestor.

Goate narrowed down the gene's likely location on chromosome 14, and though she heard a rumor that another researcher, the Toronto-based geneticist Peter St George-Hyslop, was on the same path, she forged ahead with her analysis. Both Goate and St George-Hyslop were working with Alzheimer's families from all over the world who were suspected to have mutations on chromosome 14. St George-Hyslop was using data from families in Italy, Mexico, the United States, and Canada; Goate had families in Sweden, the United Kingdom, the United States, and Colombia. But Colombia provided the lion's share of her samples, more than all her other families combined.

What happened that summer was something the Colombian investigators, Lopera especially, would remember with anguish. The two teams—Goate's and St George-Hyslop's—each found the gene independently, with St George-Hyslop beating them to publication.

In a paper published on June 29, 1995, in the journal *Nature*, St George-Hyslop and his colleagues identified five different Alzheimer's-causing mutations on the gene across seven families, the biggest of which was the family in Italy. Goate's paper with Lopera, published that October in *Nature Genetics*, described the structure of presenilin-1 in more detail, along with six more novel mutations that caused early Alzheimer's, including one unique to the Colombian kindred. "The Colombian family is a little buried in that paper," said Kosik. "But they were the largest, and that is what gave the paper the statistical power it had. And it meant that the Italian family wasn't a fluke." Still, it wasn't as glamorous to confirm someone else's discovery as to announce one's own, and Lopera, who generally spoke of Kosik as though he were a brother, was still bitter enough about the affair decades later to blame him for being scooped. Kosik "wasted precious time," Lopera said, by poking around on APP for so long before recruiting Alison Goate.

"I know he thinks that we dawdled," Kosik said. "It's true we had more than enough samples to be the first to find presenilin, and the knowledge was all there. Maybe we could have moved faster." But between the damaged samples and the error-ridden genealogies, there was a lot to straighten out first, he said. "Remember, we were doing research in Colombia at a time when half the time there wasn't good refrigeration. To go and ask a person who is a distinguished geneticist in her own right, like Alison, to use her lab's resources required some diplomacy, and she had to come to her own conclusion that this was a worthwhile project."

In 1995, the greatest scientific contributions of the Colombians were still decades away. The knowledge and technology needed to bring them about had not yet arrived. But a huge door had opened for Lopera. This extended family would launch careers' worth of studies for him and his team, and for the generation of investigators who studied under them. It would make Lopera a fixture of global Alzheimer's research and, once drug companies got involved, make him wealthy. In not being first to announce the presenilin gene, "we were obviously disappointed," Goate told me. "But I think the Colombian kindreds have really enriched our understanding of Alzheimer's in many ways beyond the mutation on presenilin-1."

SIX

No longer a mystery variant on a mystery gene, the Colombian mutation had a name: E280A. It is what's known as a missense mutation, a hiccup in DNA that causes one amino acid to be swapped for another and can change the resulting protein. In this case, the letters "E" and "A" refer to glutamic acid and alanine, the amino acids that are switched, and the number "280" describes the spot on the presenilin-1 gene where the change occurs. In later years, Lopera, when delivering one of his folksy talks to the families, likened E280A to a misspelling, an instruction gone awry: the word *mesa*, or "table," turns into *misa*, "Mass." The Colombian press dubbed E280A "the paisa mutation," because as far as anyone knew, it occurred exclusively in Antioquia, or in people who could trace their roots to Antioquia.

The investigators' priority was twofold. In order to truly grasp the natural history of the disease and understand its effects at all stages, they needed to determine which currently healthy people in the families were mutation carriers and mount a long-term study to follow them. The second concern was to see how E280A acted in the brain. Lopera urgently needed more brains from people who had died with it and to study them using immunohistochemistry, which showed researchers what disease proteins were active and how they spread.

Getting the brains was the main challenge. The group had run into enough trouble with its first. "If you show too much interest in the brains, it's a little macabre," Lopera conceded.

He soon came up with a roundabout, but inspired, solution: to announce a new brain bank at the University of Antioquia. Brain banks were an established component of neurodegenerative disease research. McLean Hospital in Boston, where Ken Kosik once worked, had been storing and studying autopsy brains since the 1970s. Lopera's bank existed only on paper; the one brain his group had succeeded in obtaining had gone off to Harvard. But he did not get into the details when he set up a kiosk at a local education fair and passed out refrigerator magnets that read: "I want to donate my brain to science." Three hundred people signed up to become donors, most of them college students who received a "brain donor" card in return. "That resulted in a writeup in the newspaper saying there's a brain bank at the university, when we didn't even have a brain yet," he said. Lopera's goal was not that healthy college kids should die, but to condition the public to think that donating brains was normal. More than anything, he wanted that message to reach the Alzheimer's families. Soon enough, Lopera said, "the families started asking us about it. They brought up the topic with us, about whether when Dad died, they should make a point of donating the brain." Lopera added a further incentive: he started paying families for donations. He framed the subsidy as a way to compensate for the inconvenience of making a donation, which forced family members to sign consent papers as quickly as possible, day or night, in their hour of grief, and to deal with the logistics of transporting a body.

Lopera and Lucía Madrigal attended their patients' funerals in those early days, partly in sympathy, partly to help fill holes in their genealogies. Wakes, which often took place in the family home, "were where we had the best opportunity to talk to everybody, because the whole family was there," Lopera said. "We'd say our Our Fathers and then go into the kitchen where everyone was gathered around the coffee and say 'Get me so and so . . . '" At one wake in Medellín, "I was talking with the family, and in the course of the conversation they said, 'Now we wish we would have donated the brain. Do you think we could still do it?' We ended up removing the body from the wake, taking it to a funeral home to harvest the brain, and then returning it. We only did that once," Lopera recalled, shaking his head. "Shameless."

BY THE END of 1995, the group had collected four more brains from Alzheimer's patients with E280A, enough to mount a study. Cynthia Lemere, a postdoctoral fellow who worked with Kosik and Dennis Selkoe at Harvard, flew to Medellín carrying a duffel full of antibodies and supplies. Lemere stayed a week, helping set up an immunohistochemistry lab. She shuttled between the San Vicente campus and her hotel, where every night an armed guard, sent by Lopera, stood watch outside her room. "I was never out of anyone's sight, ever," she said.

Lemere worked alongside Juan Carlos Arango, with whom she had studied the first autopsy brain, Florelba's, earlier that year at Harvard. Lemere showed his technicians how to stain preserved brain tissue with antibodies to disease proteins. She also accompanied Lopera on his rounds, meeting patients. "One really struck me because he was only forty-three years old. He had tried to walk to the clinic by himself and got lost, and was three hours late," she recalled. Lemere was a lab person, someone who knew early-onset Alzheimer's mainly through tissue slides and scientific papers. "It really hits home when you see that face-to-face," she said.

Arango sent Lemere back to Boston with sixteen blocks of paraffin-preserved brain tissue. "It was Christmastime, and I wrapped them up as presents," she said. She realized when she got to the airport how reckless this was. "So many Colombians got extra scans and pat-downs in the U.S. back then that there was a sort of retribution toward American passengers." Arango, after dropping Lemere off, watched anxiously through a glass window as agents pulled the wrapped gifts from her luggage. But they replaced the gifts without opening them and let her board the plane.

At Harvard, Lemere worked on the E280A brains using antibodies to amyloid-beta, which occurs in different forms, or species—some have longer chains of amino acids, and some are shorter. She took photos of her slides through a microscope and analyzed the images by computer, comparing the results with brains from older people with Alzheimer's disease. The Colombian brains turned out to have massive deposits of a type of amyloid called Aß42, which aggregates readily into plaques, while the brains from the older patients saw less of that form. Aß42 built up in the blood vessels of the E280A brains and leaked precipitously out of them. Amyloid plaques also invaded the cerebellum, the lower rear part of the brain—something that didn't happen in the older Alzheimer's patients, even when they had been sick longer.

In October 1996, Lemere and Arango published a paper showing that the E280A mutation caused the brain to become flooded with Aß42. They assumed—and it was later confirmed—that other presenilin mutations did the same. This was strong evidence to support the amyloid hypothesis of Alzheimer's disease: that overproduction of amyloid-beta, in this case caused by the mutation, triggered a cascade of harmful events in the brain.

OFELIA'S BRAIN DID NOT make it into that study. She died in August 1996, between the paper's acceptance and its publication. Ofelia was forty-four years old, making her one of the younger patients with E280A to die. She had experienced a rapid decline; it had been only three years since she lost her job as a messenger. Ofelia was from the C2 family—the clan that, even more than the Pulgaríns of Belmira, had broken the study wide open, thanks to their vast numbers and their willingness to be studied. Lopera and Madrigal attended Ofelia's funeral; later, when one of her sisters memorialized Ofelia with a short book on her struggle, Lopera addressed the dead woman in a prologue that was unusually personal for him:

> For me, it is an honor and an obligation to write this, as a physician and investigator of your story, especially the period of your illness, your death, and what will come after your death, as we conserve and continue to study fragments of your brain.
>
> I could never accept that you lost your job and your pension because no one suspected that your inefficiency, lapses, and irresponsibility were the way this scourge of dementia expressed itself in your body and mind. It was lack of knowledge of this disease that permitted society to mistreat you, in the same way it allowed your father to end up on the street, wandering and disoriented, as an expression of the same illness. Nor did your grandfather's early and progressive dementia serve to warn you of that which would be mirrored so cruelly in you.

Most painful of all, he wrote, was that "a thousand Ofelias will repeat this story in the coming years."

NOW THAT THE INVESTIGATORS could test for the paisa mutation in healthy people, Lopera wanted to explore the idea of letting people know whether they carried it. He invited Ken Kosik to come to

Colombia and help him sort it out. With Huntington's disease, for which a test had recently become available in the United States and Europe, only a quarter of at-risk people were opting to get tested. Would the Colombian families want to know whether they carried E280A?

Kosik, Lopera, and Madrigal traveled to Manrique, a hillside neighborhood in Medellín plagued in those days by endless gang wars. A young woman named Carmen answered the door for the researchers; just inside was her mother, bedridden with Alzheimer's, her body contorted and her eyes unfocused, a feeding tube affixed in her navel. Over the noise of motorcycles outside, Kosik asked Carmen and her brothers what they would do differently if a genetic test confirmed whether they'd inherited the Alzheimer's gene from their mother. At first none could really say, though they all wanted the test. Before the group left, however, Carmen's brother, who was twenty-three, confided to Madrigal that he would put a gun to his head—he made the trigger-pulling gesture with his finger—if his test were positive.

In other families he visited, rural and urban, Kosik noticed the same patterns of life: a painting of Jesus in every house. The house simply constructed and bare inside. Three generations living together, sometimes four. Alzheimer's and murder striking in nearly equal measure: *This brother had a motorcycle stolen from him and when the brother attempted to get it back from the thieves, they killed him,* he wrote of one family in his diary. *Another brother went to the hospital to identify the body, and they killed this brother as well on his way to the hospital.* And there was one more recurrent finding: every one of the at-risk people he spoke to, those who were caring for a parent with Alzheimer's or already had buried one, said they wanted to know whether they carried the mutation.

From the safety of Lopera's condominium, with its terraces overlooking Medellín, Kosik rose before dawn, a brief interlude of violet light and silence before the city's daytime clamor, and scrawled the last diary entry of his trip.

Soon there would be a test, he wrote, that *will tell Carmen and her family and this enormous extended kindred in Antioquia whether they will get the disease. They're very mortal and fragile souls reaching for divine knowledge, the presence of a mutation in their DNA that can predict the future.*

But this was not to be. The testing materialized, but only the investigators would learn the results. A few years after his visit, Kosik explained his and Lopera's thinking in an article for a popular science magazine. He cited the specter of suicide, recalling the young man's threat, and a dearth of genetic counselors in Colombia. And "so, adhering to well-established ethical guidelines," he wrote, "Lopera and I have decided that the gene test will be given for research purposes only: the results will not be made available to the person tested or to anyone else."

WHEN I FIRST MET Lopera, I asked him why his subjects were not offered the results of their genetic tests. With no cure for Alzheimer's and no effective treatments, knowledge of one's genes was the only weapon one had in confronting an uncertain future, as I saw it.

By then Ofelia's children were in their thirties and forties. One of Ofelia's daughters was enrolled in Lopera's clinical trial, and her older brother had already died of Alzheimer's, his brain stored in the same bank of freezers as his mother's. Two generations had passed without any of Lopera's subjects learning the truth about their genes—except for a few who had taken matters into their own hands and gotten tested, usually abroad.

"You're not supposed to give out data from an investigation,"

Lopera said, sounding defensive. "We're rigid on that point, and we don't care what people think." The only way to disclose genetic results properly, he said, was through a counseling program in which each at-risk person met with psychologists, neurologists, and others in a series of appointments over months. And with such potentially life-changing information at stake, you had to take multiple samples, sometimes processing them in more than one lab, to guarantee that they received the correct result. All this came with costs, Lopera pointed out.

Lopera was used to working on a shoestring. He had held his cohort together for as long as he had by putting research first. But members of his staff, some of whom disagreed with him on the issue of genetic disclosure, insisted that there was more to Lopera's reticence than concerns about cost. For more than twenty years, they said, he had allowed no formal discussion of the matter.

I asked Kosik whether he had ever brought up the topic again.

"I was their guest. It wasn't my culture. It wasn't my decision," he told me. "It wasn't like disclosure and testing was something that I was on any kind of ideological high horse about. I never foisted my opinion on almost anything about the way they did things there. I was really just glad to be part of what they were doing."

IN 1997, LOPERA, Kosik, and their colleagues published a paper in the *Journal of the American Medical Association*. Combining the team's genetic, clinical, and pathology findings from the Colombian families, it was the most comprehensive look yet at Alzheimer's disease caused by E280A.

The mean age of onset remained forty-seven, Lopera and Kosik reported. In the years before they began to experience memory loss or confusion, people with E280A reported a peculiar symptom: severe pain in the back of the head. Three-quarters of patients complained of these headaches, while in a group of control subjects—people without the mutation—fewer than a fifth did.

The paper also described the patients. The vast majority had never finished grade school. Most of the men were farmers, most of the women housewives. A surprising number had been admitted to Antioquia's public mental hospital at some point during their illness.

This last finding struck Lopera as curious enough that he tasked Julián Calle, a psychiatry resident who had just wrapped up his año rural, with digging through the mental hospital's archives. Lopera suspected that most patients went in early in the course of their illness, when a severe depression or some change in behavior preceded symptoms of dementia. At the time, and perhaps even now, psychiatrists were more likely than neurologists to see a dementia patient first.

The hospital was erected in 1878 as the Hospital Para Locos, a name soon prudently changed to the Manicomio Departmental, or state asylum. In 1958, it moved to a new building on the outskirts of the city to become what it is today: the Hospital Mental de Antioquia. Lopera's records on some 160 patients with Alzheimer's caused by E280A showed that more than a tenth of them had passed through *el mental*, as people called the institution. But what they had been diagnosed with, and how they were treated, were mysteries.

In its early days the *mental* was "your basic nuthouse," Calle explained to me between patients at the high-volume clinic where he worked. The old hospital was beautifully constructed, with ornate iron gates, stone passageways, and palm-shaded gardens. But it was a place, he went on, "where you just go and leave your family member there, and you don't know if he's ever coming back again. Because at the time we didn't have any medications."

When someone in a farming village began to behave strangely, "people would say, 'We need to take him to the *mental*,'" Calle said. For those families, the hospital offered a humane alternative to the unspeakable. "Did you read *One Hundred Years of Solitude*?" he asked. "If you remember, they chained José Arcadio Buendía"—the family patriarch, reduced in his dotage to hallucinating and muttering in Latin—"to a chestnut tree. That's what happened in rural areas."

In his search through the archives, Calle located records for nineteen patients in Lopera's genealogies. The patients had been admitted from the 1940s through the 1970s, and had many of the characteristics the researchers had come to expect in the E280A families. They were people with little education, nearly a fifth of them illiterate, with rural roots. Half had experienced symptoms for at least a year before being admitted, meaning that their loved ones had tried to care for them until they realized they could not. The majority were women, suggesting that men were less able or willing to care for wives, sisters, or mothers with dementia than the other way around.

Their diagnoses varied wildly, Calle found. A twenty-nine-year-old woman was deemed to have "epileptic psychosis," and "a syndrome of hallucinatory mental confusion." A forty-six-year-old woman was initially admitted for depression, only to be later diagnosed with mental retardation and epilepsy. Three patients were said to be afflicted with schizophrenia, three with an "organic mental syndrome," and one with drug-related psychosis. Though most had memory loss and language difficulties listed among their symptoms, only six were ever diagnosed with dementia. Not once did the hospital staff sketch out a family tree, looking for a pattern of inheritance.

The young age of some patients, admitted in their twenties and thirties, was a finding of great interest to Lopera, who along with Kosik had wondered about the possibility of homozygotes—people who had inherited the mutation from both parents—among the E280A families. Homozygotes were presumed to have it worse off, with earlier onset or a more aggressive disease course. Calle identified a thirty-year-old man who had come through the mental hospital with a depression that quickly gave way to dementia. Lopera later found that both the man's parents had early-onset dementia, meaning that he may have carried two copies of the mutation. What was noteworthy was that such a person could survive to adulthood. "We weren't even sure it was compatible with life," Lopera said.

SEVEN

The discovery of the paisa mutation made Francisco Lopera famous in Colombia. When teams from Neurociencias arrived in the hill towns, people flocked to meet *el doctor*. Lopera was a natural boss, decisive and confident, with a way of marshaling researchers that allowed them to make their own names while adding to the prestige of his. Many of the investigators who worked with him in the 1990s remember having to jockey for his favor, and also how exhilarating it was to be part of his Alzheimer's studies. It was a time when Lopera and Jorge Ossa welcomed collaborators from all different fields: "Anyone with an idea or a question we thought was interesting," Lopera said.

From a medical standpoint, there was little need to dig further into the origins of the E280A mutation. The team's goals were to keep harvesting brains, identify more branches of the family, and recruit carriers into a long-term study to observe how the disease evolved. In the long run, of course, the hope was that findings from the E280A families could lead to a drug or a vaccine that could blunt the mutation's effects. But the researchers remained curious about its history as well. Alison Goate's genetic studies had established that the families had a common ancestor. The researchers were now eager to learn who that first mutation carrier was. Mauricio Arcos-Burgos, the geneticist who worked closely with Lopera, felt confident that he or she had brought the mutation from Spain.

It was not possible yet to identify the ancestral background of E280A using genetics. The extensive reference libraries of human genomes needed for such studies were not yet available. But there were other questions that might be answered through archival work and fieldwork. How and when had E280A arrived in all these country towns? How many more branches were left to discover? Was there a common trunk to the E280A tree, a way to link all the branches?

Nearly every month in 1997 and 1998, groups from Neurociencias struck out for the villages, staying days at a time. Escaping Medellín and its relentless, enervating violence, the groups enjoyed these excursions—staying in rustic hotels, and drinking *aguardiente*, the local anise liquor, in country dance halls. Lopera acquired an official car for the "commissions," as they called the trips, an SUV with the university insignia. The commissions usually included a physician, a nurse, and a psychologist, along with a rotating group of students from a range of disciplines, including the humanities. Together they drove into the hills, chasing families still little known to them and assessing patients in their homes.

What they saw in the hamlets could be shocking. When Sonia Moreno, a neuropsychologist, and Margarita Giraldo, a neurology resident, traveled to one farm, they discovered a woman with advanced dementia living in a pigsty. "Found in deplorable conditions," they wrote in their report to Lopera. The woman, they wrote, "had food on her face, was covered in urine and feces, and was sleeping in a windowless structure made of bamboo." Some twenty-five years later, Giraldo and Moreno were still with Neurociencias, and neither had forgotten the sight. "Her face is burned into my memory," Giraldo told me. "Her own children had done this to her," Moreno recalled.

The women did not know then whether to report what they had

seen to authorities. Lopera advised them not to. "We saw some very dramatic cases of abuse and mistreatment," Lopera told me: people tied up; disabled children left helpless and abandoned when the mothers who cared for them got sick. But he didn't like to judge rural people, and if word got out that his investigators were reporting the families, he knew it could be a death knell for their research.

The investigators were finding more branches of the families. Some were as small as fifty or a hundred people, living in a single village or hamlet, while others counted a thousand or more. They had no idea how many more they might discover. "There were lots of us, working hard and working for free," recalled Margarita Ayora, who was pursuing a degree in anthropology when she joined the team. "We were struggling to piece together this whole web of unknowns."

Working alongside Lucía Madrigal, who was building out the genealogies, Ayora's job was to observe how people lived. These were extended farming families, "with parents, grandparents in the same house, who did the same things generation after generation, people with very little education but strong Christian values," she recalled. People tended to be cautious in talking about the illness, Ayora noticed, as if to avoid calling it an illness at all. As Ayora and Madrigal sat sipping *aguapanela*, the hot drink of raw cane sugar ubiquitous in country homes, a family member would describe symptom after symptom in detail, Ayora recalled: "'He had a lost look. He lost things, he lost money.' This was a constant theme with them, losing money. Or, 'I had just served him his lunch and he got up furious demanding to know where his lunch was when he'd just eaten it.'" It was common to call someone *necio*—childlike, obstinate—as a way of describing their behavior. Or *bobito*, silly. But they did not call the person sick.

Did they think the disease was caused by witchcraft? I asked Ayora.

"Sometimes. People who believed 'they gave him herbs,' especially some woman," she said. When confronted with the name and symptoms of the disease—Alzheimer's—"they said they would go ask Padre Marianito to cure them."

Mauricio Arcos-Burgos's insistence that the E280A family had a Spanish founder seemed to square with what Ayora and the others were seeing: farming families, most of them white, at the dawn of the twenty-first century, still living like colonists. But to identify a common ancestor of the E280A families, if such a feat were even possible, would require a systematic review of church records and Spanish colonial archives—an exhausting process that might take years. The earliest case Madrigal could identify, through interviews with elders, was a man named Mauricio Pineda Sampedro, a patriarch of the C2 family said to have died in the early nineteenth century. Madrigal had experience working with church records, and she had notebooks filled with years' worth of clues, but she could not dedicate herself full time to a project of purely historical interest. The task of searching for the origins of E280A fell to Liliana Cadavid, a journalism student at the university who wrote for one of its magazines.

Cadavid signed up with Lopera for a small salary and the allure of a mystery to solve. On the surface, her task was simple: locate dates of birth and death for the people named in the genealogies. Untrained in archival research, she started off with her backpack, her camera, a change of clothes, and a blank notebook for Angostura, the hill town in whose outlying hamlets the C2 family, the largest of all the E280A families, was concentrated. Cadavid discovered that Mauricio Pineda Sampedro was born in Medellín in 1793, to a couple named José Miguel Pineda and Petronila Sampedro. Mauricio's descendants, some of whom settled in Angostura, had died in midlife with the chilling frequency of E280A. But neither Mauricio nor his wife could have been the first person to carry the mutation, because during their lifetimes, Cadavid saw, the disease was already occurring in other Antioquian towns. Cadavid's notebooks piled up, full of records she had hand-copied from parishes all over Antioquia, but the common ancestor of the E280A mutation continued to elude her.

It was not until nearly two years later, in a town called San Andrés de Cuerquia, that Cadavid discovered two women with the surname Sampedro: Mauricio's maternal aunts. In time she would find four Sampedro sisters: Petronila, Francisca, Luz, and Modesta. They had lived in the late eighteenth and early nineteenth centuries, and all had married and raised families in different Antioquian towns.

The towns were of a piece: cool highlands rich in fresh waters. The era in which the sisters lived followed a rush of migration from Antioquia's lowland shaft mining towns to its highlands, where gold could be mined in streams and fertile soils fed a rapidly growing population. The Sampedro women and their descendants moved around quite a bit. But they moved as though within an ecosystem, a terrain they knew how to work and exploit. For some two hundred years they remained in the hills, farming and reproducing and falling ill and dying without the outside world taking notice.

THE SAME WEEKEND THAT she discovered the Sampedro sisters, in the parish archives of San Andrés de Cuerquia, Liliana Cadavid narrowly averted being dragged off a bus by right-wing paramilitaries, who were suspicious of her having arrived in town with notebooks filled with names. With the bus stopped and its passengers staring in horror, she struggled to explain to a man carrying a rifle that the names were of people who had been dead for centuries. Cadavid did not return to the field, and in 2001, she left Colombia for the United States.

More than a decade later, Cadavid contributed her data, some of it still stored as handwritten records in composition notebooks, to support Ken Kosik's molecular studies of E280A's origins. Using samples from some seventy-five E280A carriers, Kosik and his colleagues determined that the E280A mutation occurred on a haplotype known to have originated in western Europe. The scientists were also able to give the mutation an approximate age. They estimated that E280A was fifteen generations old, meaning that it likely emerged in the first half of the seventeenth century.

Kosik had always favored the idea that E280A had arrived in the New World with a conquistador and later died out in its native Spain. But the mutation had never been reported in Spain, and its first carrier could have been anyone with European ancestry: a recently arrived colonist, a Colombian-born person, a man, or a woman. There was no way to know. "The bigger question is fixation," Kosik said. "That is, the interaction with the environment that caused the mutation to become fixed in the population to the degree that it did."

Cadavid, in discovering the Sampedro sisters, had found the answer.

"We never thought—it never occurred to us—that it was women," she told me. The image of the male conquistador, or colonist, was so ingrained in the researchers' minds that it all but blinded them to the possibility. Mauricio Arcos-Burgos had envisioned an itinerant Spanish-born man, perhaps a miner, as the "patient zero" who had brought E280A to all these towns. Instead, here were four women whose fecundity allowed the mutation to take hold quickly, across vast terrain, in a single generation. Tallying the descendants of Petronila, Francisca, Luz, and Modesta, Cadavid counted more than five thousand people.

She liked to picture the sisters as grandmothers, robust matriarchs of the mountains. They may have been grandmothers when they died, but they would not have been old; each would likely have developed dementia in her forties. Were the Sampedro sisters cared for by their children, carefully fed and laid out on wooden beds, watched over by phalanxes of crucifixes and saints? Or were they subjected to the kind of pitiful neglect that the investigators would uncover centuries later?

BY THE EARLY 2000S, the Colombian conflict had heated up to the point where the risks of fieldwork had come to outweigh the rewards. Regional paramilitaries were coalescing into aggressive, invading armies, and the guerillas responded with more desperation and more violence.

Lopera and his investigators stopped traveling to the hill towns. "It didn't seem wise to continue exposing ourselves, especially when we had so few resources," he said. The city offered only partial refuge from the conflict; beyond the hospital walls, urban guerilla cells battled paramilitaries in the comunas of Medellín. One morning the Neurociencias team found that their vehicle had been stolen from a university lot; it was later discovered in a guerilla stronghold three hundred miles away.

Lucía Madrigal left Colombia for Spain to pursue a doctoral degree. Margarita Ayora left the group soon afterward, and Alison Goate wrapped up her genetic studies. With fewer research funds on hand and no way to get out to the towns, Lopera focused on conducting a long-term natural history study—a type of study to see how a disease develops—from his clinic in Medellín. He began following the healthy sons and daughters of people who had become ill, knowing that half of them, too, would go on to develop symptoms whose progression could be tracked until they died. Some family members obliged, allowing themselves to be tested for the mutation and observed over time, while others drifted away. It was an uncertain moment for Neurociencias, a time when no one knew whether the Alzheimer's studies would continue to bear fruit. Participating in drug research was a remote, almost abstract, prospect.

In the early 1980s, when Lopera started his residency in clinical neurology, he knew he was making an unpopular choice. A neurologist was seen as someone condemned for life to treat incurable diseases: "to pour water on vegetables," as one of Lopera's professors had cruelly put it. But in the early 2000s, this was changing, just a bit. Genetic and brain pathology studies pointed to many previously unknown pathways at work in neurodegenerative diseases like Huntington's, Parkinson's, and Alzheimer's. Still, therapies based on these discoveries remained out of reach. The first drugs to treat Alzheimer's disease, which appeared in the 1990s, were modeled on previous treatments for Parkinson's and targeted the neurotransmitter acetylcholine. The drugs helped people stay sharper for a spell, but their neurons went on to die anyway. Lopera was still pouring water on vegetables.

He sought grants to study other diseases. He accepted speaking invitations from Rotary Club chapters and the Medellín city council to recount his achievements with the paisa mutation. He memorialized his experiences in a long essay titled "La peste de la memoria," the plague of memory, an allusion to Gabriel García Márquez's mysterious forgetting disease. In it he described his first impressions of Pedro Julio Pulgarín, in 1984, as a man "struck by the memory plague of Macondo," one whose gaze, Lopera wrote, "did not focus on people or on things and meandered as if probing infinity. Pedro Julio's family sought only relief for his delusional and paranoid ideas, as though they had already accepted with resignation that he was merely condemned to repeat an ancestral destiny."

EIGHT

Most Alzheimer's researchers felt that a successful disease-modifying therapy would have to target amyloid-beta. Because amyloid was the first protein seen emerging in the disease, and because genes like APP and presenilin acted to increase it, it was held to be the foundation of the complex process that followed. If an intervention could be given before amyloid plaques had a chance to build up, Alzheimer's might be prevented; if it was given later, the disease could perhaps be slowed. One idea was using drugs targeting the enzymes needed for amyloid-beta to form, while another was a vaccine that prompted the immune system to attack amyloid and clear it from the brain. Large pharmaceutical companies had yet to dedicate real resources to designing anti-amyloid therapies, leaving small biotech firms to try their luck.

In 1999, Dale Schenk, a researcher at a San Francisco startup company called Elan Pharmaceuticals, injected synthetic amyloid-beta into mice that had been genetically engineered to develop excess amyloid and found that plaques did not accumulate in their brains. Schenk's discovery earned him huge public fanfare, but some researchers were wary, arguing that no mouse model could capture the complexity of Alzheimer's in humans, and it was still not established whether amyloid plaques were a cause of the disease or a byproduct of it.

In 2000, Schenk tried the same approach in people in the early stages of Alzheimer's disease, with disastrous results. The vaccine cleared plaques, but it also sparked an inflammatory reaction in some patients' brains. A follow-up study showed that patients whose amyloid plaques had been cleared by the vaccine saw their disease progress anyway, and they died with advanced dementia. But the dream of an Alzheimer's vaccine persisted. As researchers pondered what had gone wrong with Schenk's vaccine, some thought that timing may have been part of the problem: most of the people in the trial already had symptoms. Once the disease cascade was well in motion, they reasoned, it was possible that no drug or vaccine could have helped.

But what if you selected people genetically destined to become Alzheimer's patients and treated them before symptoms emerged? Could the disease be stalled or averted? Though researchers had identified common genetic variants that increased the risk of Alzheimer's disease, these were not as certain to cause the disease as the early-onset genes. Presenilin mutations did not merely raise one's risk, the way having a copy of the APOE4 genetic variant did; they all but sealed one's fate.

It was still unclear, however, whether people with presenilin mutations offered a good model to study drugs for Alzheimer's disease. The ideal therapy was one that would treat not just the rare familial early-onset forms, but the much more common late-onset disease as well. Some researchers nonetheless argued for testing in people with presenilin mutations, making an analogy to statin drugs, which were first tested by a Japanese scientist in people with mutations that caused excessive cholesterol buildup. These carriers often suffered from coronary artery disease—and sometimes fatal heart attacks—in their twenties. Eventually, of course, statins proved successful in the general population as a treatment for high cholesterol. The rare familial form of the disease turned out to share a biological pathway with the common one.

AT THE INTERNATIONAL CONFERENCES he attended in those years, Francisco Lopera was intrigued by talk of an experimental Alzheimer's vaccine. He felt that healthy carriers of E280A would be especially well suited—more than Dale Schenk's subjects, who were already ill—to participate in a clinical trial of a preventive treatment. A well-designed vaccine might even stave off their disease indefinitely, he thought.

Ken Kosik noticed that on his visits to Colombia, word of an Alzheimer's vaccine was circulating not just among the researchers but among the E280A families as well. After years of taking part in study after study without seeing much benefit beyond a day's wages and lunch, many had lost interest in Lopera's research program. But when Lopera began talking about a vaccine, they were suddenly full of questions again. "He put the bug in their ears," Kosik said.

Large pharmaceutical companies were gradually becoming intrigued with the concept as well. In 2006, after years with little outside funding for his Alzheimer's research, Lopera received a call from the Swiss drug maker Novartis, which was developing its own, ostensibly safer version of an Alzheimer's vaccine. Lopera's contact at Novartis, a Mexican-born researcher, proposed building a formal registry of E280A carriers and noncarriers whose blood, spinal fluid, and brain images could form the basis for a future trial. The funds he was offering to do this—around $2 million—were unlike anything Lopera had dealt with before. Neurociencias had recently moved to a new research complex at the University of Antioquia; until then, the investigators had worked out of a rental house whose computers had been rescued from the university junk closet. Lopera flew to Basel, where he was well received by the Novartis executives. For two years they went forward building a registry. But there was something that bothered Lopera about the way the company aimed to conduct its research; it was as though Novartis aimed merely to contract a service from him. A draft contract submitted to him to review made it clear that Novartis "wanted to own everything," Lopera recalled—including all his data. Worse, he said, it allowed Novartis to impose fines if he failed to recruit enough people for the registry. Colombia, as Lopera saw it, was a country at war; an offensive by paramilitary or guerilla forces in the hill towns could paralyze recruitment indefinitely. There was no way he would agree to a fine.

The University of Antioquia had a team dedicated to negotiating research contracts with companies, but Lopera trusted himself over anyone at his own institution. He opted to manage the deal alone, relying on the country wisdom he'd inherited from his late father, Luis Emilio Lopera, a farmer-turned-shopkeeper who had never set foot in a school. For Luis Emilio every deal was a handshake deal; that was the old Antioquian way. You were firm, your word was good, and when you didn't like the smell of something, you walked away. "The answer I eventually gave Novartis was straight from him," Lopera said. "*Con usted no hay negocio*"—there's no doing business with you.

"It was hard for someone in my position to turn down a company like Novartis," he recalled. "We needed that money." And his patients, of course, needed drugs. Lopera had long hoped for a therapy for families, perhaps one that arose from studies of their own bodies. *We will not lose the hope of encountering some clue in your frozen tissues, to advance the search for a therapeutic option effective against this scourge*, he'd written in his elegy to his late patient Ofelia.

The failure of the deal with Novartis left Lopera doubting himself. He had walked away from the kind of cash he was not sure he would ever be offered again. But the ideas percolating then in the Alzheimer's world—the new focus on preventing the disease and detecting it long before symptoms emerged—gave him reason to expect further overtures. Ken Kosik, who had recently moved his lab to the University of California, Santa Barbara, never hesitated to propose studies enrolling the E280A families to other Alzheimer's researchers, regardless of the discomfort some showed when he did. North Americans and Europeans associated Colombia mostly with drugs and mayhem, and Kosik's travel there was viewed as just another of his eccentricities, like his decision to leave a tenured position at Harvard for a state university.

Kosik had left Harvard because he'd grown tired of being surrounded by molecular cell biologists like himself in an institution where "the translational gorilla"—the pressure to direct all lines of inquiry into something of potential therapeutic use—"was always in the room," he said. Kosik didn't believe in translating anything into a medical application before the basic biology of it was absolutely buttoned down, and the biology of Alzheimer's was far from settled. In fact, a significant minority of researchers remained unpersuaded by the amyloid hypothesis, thinking instead that mutations on presenilin and other genes provoked damage in ways not mediated by amyloid-beta.

UC Santa Barbara, which sat atop cliffs overlooking the Pacific, was a universe apart from the Harvard medical campus. The marine biology department's boats dredged up strange sea creatures whose genes Kosik could study; from his office window he could sometimes see whales playing. He liked that there were whole buildings dedicated to music and theater. As a rare top biologist at Santa Barbara, Kosik knew the post would give him unusual opportunities to collaborate with researchers in other fields, especially physicists, engineers, and computer scientists. His instincts proved right, and he never regretted his decision.

He'd been in California two years when he got a call from Eric Reiman, an Alzheimer's researcher who had just set up a new institute in Phoenix, Arizona, around the novel idea of Alzheimer's prevention studies. Reiman, a psychiatrist by training and an expert in brain imaging, studied people with variations on a specific genetic risk profile for late-onset Alzheimer's disease: those who had one copy, no copies, or two copies of the APOE4 variant. Reiman had found that years or even decades before they were likely to develop dementia, people with two copies of APOE4—a genetic profile shared by about one in fifty Americans—saw changes in how their brains metabolized glucose, the brain's principal source of energy. Reiman was beginning to see Alzheimer's not so much as synonymous with dementia but as a biological continuum that began earlier in life than anyone had thought.

At first Reiman didn't consider testing drugs or vaccines; his intention was simply to follow healthy people over time, to see what factors—including cholesterol, weight, and blood pressure, as well as APOE variants—influenced their getting the disease. "But then I asked myself how many healthy people, and how long it would take," Reiman said, "and I estimated it would take fifty thousand people in a twenty-year trial." This was not something that a pharmaceutical company would want to subsidize, or that donors would have the patience to underwrite.

Reiman wondered whether, instead of waiting to see which at-risk people developed dementia, you could instead look at the preclinical changes occurring in people who were destined to become sick: markers of amyloid and tau in cerebrospinal fluid, brain images, subtle cognitive tests that marked problems long before memory loss kicked in. These, Reiman proposed, could serve as indicators of disease progression. Drugs targeting amyloid were not yet ready to be tested in people at risk for Alzheimer's. But one day soon they would be, and the idea was to be ready for when the moment came. Reiman had known for years about Kosik's work with Lopera, and he knew that the E280A family was estimated to comprise five thousand living members, of whom at least a few hundred would be of the right age to participate in a trial.

Alzheimer's disease was a favorite cause among philanthropists in the United States, and Reiman had a gift for fundraising. His nascent organization, the Banner Alzheimer's Institute, sought funds from private donors, drug companies, and government agencies alike. Reiman's sudden interest in Lopera's families made a Colombia project seem like a reality, Kosik said, "compared to the castle-in-the-air dreams Lopera and I had been discussing for years." Kosik did have misgivings about the science behind any amyloid-clearing therapy, and they would only grow with time. But like Lopera he also knew firsthand the desperation of the Colombian families. If one drug failed in Colombia, at least they would have established a bulwark, laying down the infrastructure to host further trials down the road. Reiman was offering the chance to build that, with a promise of $1 million to start.

Lopera flew to Arizona, where he was received by Reiman and his colleague Pierre Tariot, the codirector of Banner Alzheimer's Institute. Reiman was an imposing figure: a tall, loquacious man who spoke in a corporate patois full of "thought leaders" and "stakeholders" and was adept at steering any conversation in the direction he wanted it to go. Tariot was reserved and pensive, a literature and classical music buff who had also come into Alzheimer's research by way of psychiatry. Lopera liked the idea of working with this small Arizona research group more than with a drug company—"things felt a little more equal," he said—and his trusted colleague Ken Kosik had more faith in Banner than in Novartis.

VALLEY OF FORGETTING

THE BANNER SCIENTISTS were aware of the potential pitfalls of a long-running Alzheimer's drug trial in Colombia. Thus far, they knew, Lopera had studied only a few hundred carriers of E280A, and estimated that there were at least five thousand more living family members out there. Reiman and Tariot wanted to know if those people could be found and tested. Was the population's education level so low that it would affect cognitive tests? Would Catholic women practice strict birth control during a study that could go on for years? Did many people abuse drugs and alcohol, or use herbal medicines that could alter their cognition? The Banner researchers knew little about Colombia, and what they'd been told, from a Medellín-born surgeon they worked with in the United States, was discouraging. Colombia was too dangerous, the surgeon insisted. They'd never be able to recruit patients under the current conditions there; patients could never withstand the rigors of a long-term study; and few in Colombia had experience with clinical trials.

On the opposite end of that argument stood Kosik, who felt that none of this was necessarily insurmountable, and that Reiman and Tariot should come down and have a look for themselves. In 2008, they did. With Lopera serving as host, they gathered family members in a university auditorium and drove to Yarumal and Angostura for meetings in community centers, chaperoned by mayors and priests, trying to explain their mission. While riding with Lopera, Reiman and Tariot were startled to see a column of men and women in fatigues and rubber boots, walking in a shallow river. Are those soldiers? Tariot asked Lopera. Maybe, Lopera answered, in a tone that discouraged further questions. The Banner scientists had gotten a little taste of what he and his colleagues had been dealing with for years.

Reiman and Tariot came away impressed with Lopera, and with

the families, who seemed to understand at least the basics of what the scientists were proposing. At every meeting, Tariot pressed the families to tell him what they hoped to get out of participating in a drug trial besides the obvious thing, a cure. One simple, consistent response he received surprised him: adult diapers. In the advanced stages of Alzheimer's, a patient could go through several diapers per day, and they were expensive.

Poverty made doing clinical research in this part of the world automatically fraught. Generous study payments were widely criticized as enticements, but medical ethicists tended to agree that something was owed to trial participants in developing countries. Some felt that participants were entitled to the study drugs indefinitely if they worked, but companies did not like to make hard promises, especially about experimental drugs that might never get regulatory approval. Others advocated for compensation in nonmonetary forms, such as better healthcare or preferential employment.

And still others proposed that bolstering the research capacity of the host country was a valid, if indirect, way to compensate a study population for contributing to science. This was how Kosik saw things, and strengthening the work of Neurociencias was one of his top priorities. One thing he hated about Nancy Wexler's studies of Huntington's disease in Venezuela was that her data—decades worth of blood, semen, tissue samples, genealogies, and interviews—went back to the States, leaving little for her Venezuelan colleagues to work with. Over the years, Kosik had been on guard that no outsider should use Colombian genes or other data without an authentic collaboration with Lopera's team.

A clinical trial, and the huge capital investments it necessitated, could vastly increase Lopera and his team's ability to study Alzheimer's disease, but only if it were set up in their favor. Lopera, who had learned to drive a hard bargain with outsiders, insisted that all data from any trial and the samples from the family members must remain available to his team. He refused to accept any role on the trial besides that of principal investigator, even if he had to share that role with Reiman and Tariot. Finally, Lopera demanded that the sponsors pick up the tab for his *plan social*, the workshops, classes, and support groups that his group had long offered to all members of the E280A families. None of this lessened their poverty, but at least it sought to assuage some of the psychological and social burdens that came with caring for someone with early Alzheimer's, or from knowing oneself to be at risk for it.

IN 2009, REIMAN and Tariot convened some forty Alzheimer's experts, including Kosik and Lopera, for a two-day meeting in Phoenix. The strategy they discussed involved recruiting for a trial in the absence of a therapy, on the expectation that something would come up within a year or two—preferably an anti-amyloid antibody. Genetically engineered monoclonal antibodies were a groundbreaking if pricey class of drugs. Used in autoimmune diseases and cancer to destroy harmful proteins, they had lately shown promise in treating a neurological disease, multiple sclerosis. Instead of prompting the body to make its own antibodies and mount its own unpredictable attack, these externally produced antibodies could be administered in controlled doses, making them theoretically safer than Dale Schenk's vaccine.

A trial enrolling E280A carriers would require that participants had no symptoms of dementia when they enrolled. But no trial could go on long enough to wait for all of them to develop dementia, especially since the age of onset was not entirely predictable with E280A. The answer to this problem came through the work of Lopera's colleagues at Neurociencias, who were discovering subtle memory glitches even among carriers in their thirties. The carriers aced standard memory tests but stumbled with what researchers called binding: remembering both the shape and the color of an image they'd just seen. Noncarriers the same age had no such difficulty. Functional brain imaging studies in these thirtysomething carriers showed that their brains had to work harder during memory tests than those of their noncarrier relatives.

By their middle thirties, all the studies suggested, E280A carriers and noncarriers began to diverge measurably in cognition, though dementia did not set in until the forties. The tests used in the drug trial would have to be sensitive enough to catch elusive signs of decline in people who were still years away from dementia.

There was still much to be hammered out. But the Phoenix meeting sealed the decision: there would be a trial in Colombia. Lopera, in typical fashion, was quiet throughout most of that two-day meeting, but he had the last word, and none of the researchers forgot it.

"The families are waiting for you," he told them.

NINE

Ledy Piedrahita and her husband, Danilo, the parents of the nurse Francisco Piedrahita, had lived most of their lives in the hamlet of Canoas, on a farm they struggled to maintain. In recent years the pair had uprooted themselves to join their adult children in the Medellín suburb of Bello. The family shared a long railroad-style apartment with polished floors and a view of a busy street, which Danilo spent much of his time staring at while smoking.

I marveled at how much Ledy and Danilo looked alike—they shared a compact, trim physique, small features, and bright eyes rimmed with dark lashes—until I remembered that they were first cousins. Ledy's married surname was Piedrahita de Piedrahita.

Ledy's father had died of early Alzheimer's, leaving her a 50 percent chance of having inherited the E280A mutation. Two of her sisters had died of it; another was just starting to show signs of the illness; a younger brother was entering full-blown dementia. Both of Danilo's parents had died of E280A. It was not unheard of among the families for both members of a couple to carry the mutation, and the researchers had identified from blood samples a half dozen people in the cohort who were born homozygotes, victims of a double hit. Two of Danilo's sisters had died of Alzheimer's while still in their thirties, suggesting that they might have been homozygotes. As a child of two carriers, Danilo's odds of having the gene—either one copy or two—were 75 percent.

And yet, for all this, Ledy and Danilo appeared to be just fine. They were in their sixties, and neither had developed symptoms. Their son Francisco, the neurology nurse, had trouble believing that they had truly been spared.

Ledy told me that she felt Francisco suffered from a kind of survivor's guilt. He thought that if his nuclear family had somehow escaped the disease, it was only by way of a miracle. Miracles exist, Ledy assured Francisco.

At her dining room table, Ledy flipped through a pink notebook. Danilo could barely sign his name, she told me. But she had made it through fifth grade, and as a grown woman she had drafted a memoir, in a tight, elegant script, of life in Canoas before and after the arrival of Lopera and his team in the 1990s. "Remembering what others have forgotten," she'd titled it first. Then she changed it to "Experiences to remember."

Ledy began reading aloud.

It was something common in our community. The people who got sick presented symptoms of madness. People took them to the doctor but they didn't know what the cause was. Their families didn't either. They were just considered crazy. Time went by. One after another became ill, until the sickness increased bit by bit. None of us worried about it.

I interrupted. Nobody worried about it?

Not in the way you would expect, Ledy said. People would not admit to worrying for themselves, but they were quick to point out symptoms in others: He lost his hat, he lost his ruana. They mocked anyone with symptoms. *Se le corrió la teja*, they would say—his roof tiles fell off. *Está picado*—he's gone bad, putrefied.

Everyone created their own myth to justify the death of their loved one: They put a spell on him. They gave him herbs. They put a curse on him. They're saying prayers. They gave him a drink. Someone came and gave it to him.

She paused to explain: "There were people who thought it was contagious." One of her brothers had believed that, she said, until Lopera persuaded him otherwise.

Ledy continued reading. *Too much work made him sick. They took blood and that made him sick. He didn't take enough vitamins and he exhausted his brain and became crazy. He got involved with a witch.*

This was what people said about Ledy's father when he got lost near Yarumal and was discovered three days later, disoriented and with hypothermia: that he'd gotten entangled with a witch. Never in his life had Ledy's father engaged in magic or spells of any kind; the Piedrahitas considered that sinful. They were devotees of the Virgin Mary, the Holy Trinity, and their local saint Padre Marianito, not witches. Everyone around them knew that. But that did not stop people from repeating the accusation.

He passed under the manzanillo tree. That was a tree whose toxic sap could cause uncontrollable itching, dangerous allergic reactions, and even death.

He went under the arboloco.

The *arboloco*? I asked. I'd heard about the manzanillo, but this was the first I'd ever heard of the arboloco, the "crazy tree." It was a name people used for a tree that once grew everywhere in the region, Ledy explained, though you didn't see so many now. Danilo, popping in from the balcony with sudden interest, explained further. The arboloco had broad leaves and multiple narrow trunks that radiated from its base, he told me. When the tree was young its wood was hollow, like bamboo, and filled with a soft white pulp.

Later I looked up the arboloco. Its scientific name, conferred in

1864 by a German botanist who'd probably never seen one alive, was *Montanoa quadrangularis*. A common tree throughout the Andes, it was used widely for building in colonial times. I found an exhaustive modern-day study of the tree by a Colombian botanist, who had analyzed its chemical properties. There was no mention of any association with madness or illness, and when I emailed the botanist to ask, he said he'd found nothing in it that could cause neurological symptoms.

Ledy thought it was probably the name of the tree, the mere fact that it contained the word *loco*, that gave rise to the association. People believed anything, she said. Even long after everyone in Canoas had been told that the sickness was a hereditary disease with a name, many people still clung to ideas about witchcraft and magic. After all, these were things you could do something about. With a disease you had to trust science, which was plodding, mysterious, and entirely the domain of others.

ONCE THE DECISION TO proceed with a clinical trial was made, the most urgent issue facing the scientists was finding enough people to enroll. By that point, in 2009, fewer than two thousand of the E280A family members had been formally studied, but the genealogies indicated that there were at least five thousand living. Lopera, who was careful never to promise more than he could deliver, gave Pierre Tariot and Eric Reiman that number, five thousand, as a target as they began work on the registry—the large pool of family members from which the scientists could collect data and select participants for a clinical trial.

The first challenge was locating them. It was one thing to know about twelve siblings and their children from names in a genealogy, quite another to track all of them down and examine them. Complicating matters was the dire state of security in the countryside; by this time the paramilitaries' influence had eased, thanks to a government push to disband them, but obscure shifts of power were occurring all over, as ex-paramilitaries formed new mafias and guerillas scrambled to reclaim territory they'd lost.

Lopera turned to Ledy Piedrahita. As her son would do years later, when he took groups to Boston for brain imaging, Ledy put people at ease. She was the type of person who, when FARC guerillas showed up to her kitchen looking for food and information, could send them away with neither and still have them thank her for her kindness.

From Canoas, Ledy started making calls to her relatives among families in "red zones"—areas under guerilla or mafia control—spreading news of the trial. Lopera had instructed Ledy not to enter these areas, but she could do her best to get people out of them and into Lopera's vans. If a family had paramilitaries or guerillas among them, she was undeterred in getting those members into the clinic, too, just mindful about with whom they would travel. At the plaza in Angostura, Ledy picked up regular faxes from the Neurociencias team with lists of names. Her task was to locate and register the people on the lists, but after a while the lists became less relevant. She was finding people the investigators had never heard of. Sometimes Ledy filled three vans in a week. One clan she took to Medellín filled nine.

By 2010, Lopera and his colleagues were feeling confident enough in their progress that they were ready to make their efforts public. They invited a *New York Times* reporter, Pam Belluck, to Colombia. The resulting article, which ran on the front page of the *Times*, featured several members of the Piedrahita clan. It also announced the scientists' plans for a trial, though there was still no drug in sight. And it conveyed a sense of the desperate stakes, for the E280A families and the world.

The next year the Banner scientists began flying members of the families to the Banner headquarters in Arizona, where they underwent positron emission tomography scans for amyloid. The PET imaging would show when and how plaques accumulated in their brains, a step toward knowing what age range of patients to include in a trial.

The groups comprised mutation carriers of every age, from healthy twenty-year-olds to sick people in their forties and fifties accompanied by caretakers, and many noncarriers for comparison. For many of the travelers, it was their first time on a plane. Ten groups of a dozen or more reached Phoenix that year. Travel is hard on dementia patients, and the sick ones suffered. One man was stopped in Miami for having the same name as a wanted criminal, and when he could not answer the officials' questions, he was harassed and detained, causing the group to miss a connecting flight. During a layover in Los Angeles, a woman tried to flee her hotel, believing she was lost in Yarumal.

In planning for the Colombians' arrival, Pierre Tariot thought about what he'd seen on his trip to Antioquia and how he might want to be received if he were a paisa overseas. He greeted every group at the airport. He hosted dinners at his home, managing to source arepas, as the families had requested. He rounded up as many of his own family members as possible, to create a welcoming atmosphere, and hired a pianist to perform for his guests. Banner staff and volunteers arranged tours of Phoenix, evenings at Mexican restaurants, trips into the desert, and outings to the zoo and to the mall, with gifts of spending money. The Colombians visited the Banner offices, where they were startled to see black-and-white pictures of their relatives, the Piedrahitas, hanging in a conference room. The photos had been taken by *The New York Times* photographer in Yarumal, Tariot explained. He'd had them blown up and mounted to remind everyone at Banner of their commitment to the families.

The scans from Phoenix revealed that amyloid plaques began to accumulate in E280A carriers' brains from as early as age twenty-eight. This finding tracked with previous work on E280A carriers' brain function, which showed a host of changes happening before even the sharpest clinicians might be able to perceive them.

The trial would enroll young people with few or no plaques, as well as people who had plaques but had yet to develop symptoms. The scientists wanted E280A carriers in their thirties and forties whose daily functioning was not yet impaired, who did not use hard drugs, who did not mind using two forms of birth control for years on end, and who were deemed likely to comply with the exacting requirements of a long-term study that included spinal taps, pregnancy tests, MRIs, and PET scans.

THE ONLY PEOPLE NEEDED for a trial like this were mutation carriers. But there was a complicating factor: Lopera had made it his policy never to disclose genetic results to anyone besides the few investigators who needed to see them.

If he recruited only mutation carriers to receive the study drug or placebo, people would realize they were carriers whether or not they desired to know. If he recruited a mixed population of carriers and noncarriers he could avert this, but at the expense of enrolling people without any reason to take part in a drawn-out, years-long trial that imposed severe restrictions on their lifestyle and reproductive choices.

In light of this unusual dilemma, the Banner scientists convened an ethics and cultural sensitivities committee to evaluate potential trial designs. Jason Karlawish, an Alzheimer's researcher at the University of Pennsylvania, was tasked with sorting them out. Karlawish had not been to Colombia, but it was his understanding that the mutation was associated with a strong social stigma there. He explained his thinking a few years later in a talk to physicians. "In Colombia, genetic testing is not the norm," Karlawish told his audience. "There are no genetic counselors in Colombia. There are none. And for this particular disease the standard in the community was not to know. Was not to find out. Why? Concerns about employment certainly, but really issues around family. In these communities, knowing that you had the disease essentially fated that you would not be able to have a family."

On this point, Karlawish was either misinformed or had been misled. Most people in the E280A cohort who were of an age to have children did, with partners who were aware of their family histories. And while genetic counselors—health workers trained to help communicate disease risk and test results—were few in Colombia, there were physician-geneticists who offered testing and advice.

Karlawish concluded that the best option was to protect the status quo by way of a three-arm trial that would allow noncarriers of the E280A mutation to participate while keeping everyone blinded to his or her genetic status. Half of the carriers would receive the study drug, the other half placebo. Every noncarrier in the study would receive placebo. The committee agreed to the proposed design—including Kosik, who had long deferred to Lopera on the issue of genetic disclosure.

AS THE FAMILIES were flying to Phoenix and back, the Banner team of Reiman, Tariot, and their colleagues entertained proposals from pharmaceutical companies. Among Lopera's demands, besides funding for his social programs and a ban on genetic disclosure, was that any drug to be tested in the E280A families had to have an immaculate safety record. This was a tall order. Lopera was mindful of the fiasco with Dale Schenk's vaccine, and more recently the first anti-amyloid antibodies tested in people with Alzheimer's had been found to cause brain swelling. It was bad enough when this happened to people who were old and sick. The thought of a thirty-five-year-old developing brain swelling or bleeding was a dealbreaker. The prospect of delivering the study drug through intravenous infusion, as most monoclonal antibodies required, did not sit well with Lopera either. To lie in a chair for hours with a line in one's arm reminded him of treatments for cancer. The trial participants may have been people destined to develop dementia, but they were still healthy, in the prime of life; in his view, they deserved to feel as normal as possible.

Finally, Lopera also had his own safety to consider. Any experimental drug carried the risk of fatality, and it was not unheard of in Colombia for doctors to be killed in retaliation for a patient's death. It had nearly happened to Lopera during his government service, when a woman died on the operating table. Lopera never forgot being called a *médico asesino*.

AFTER LOPERA MULLED and rejected several drug candidates for the trial, one emerged that fit his exacting criteria. Carole Ho, a neurologist working for the San Francisco biotech firm Genentech, introduced the scientists to crenezumab, a monoclonal antibody the company had recently acquired from a Swiss startup. The drug had been shown in healthy volunteers not to cause the brain swelling seen in previous trials. Helpfully, each study dose of crenezumab could be delivered by a quick pair of injections to the abdomen or arm.

Ho had known about the Colombian cohort since the 1990s. As a medical student on a research scholarship, she had spent two years in Kosik's Harvard lab, coming in just as Kosik was writing up the findings from the first E280A brain. Kosik was traveling to Colombia frequently then, returning full of stories and straining to improve his

Spanish. Now it was Ho's turn to spend time in Colombia, scheduling the first of multiple trips as Genentech evaluated the prospect of a vaccine trial for the E280A families.

To Ho, the Colombian trial represented a potential landmark, a better test of the amyloid hypothesis than any to date. No Alzheimer's therapy had yet been tried in people without symptoms. "This was going to be something that could really answer a major clinical scientific question," she said.

Ho's employer, Genentech, and its parent company, the Swiss pharmaceutical giant Roche, had to decide whether it was even feasible to conduct the crenezumab trial in Colombia. No Medellín hospital then had a cyclotron, the particle accelerator needed to create radioactive tracers; one would have to be acquired to conduct the PET-amyloid imaging that the trial would require. But Ho's visits were about more than making sure the university was equipped for a trial, or that the trial could enroll enough people. "It was important to me to actually see all of the sites and understand the geography of Colombia to decide—could we actually do this?" It became clear to Ho early on that they could not ask every participant to travel to Medellín. Some of the people who attended meetings in the hill towns, she saw, had to take a horse, a moped, and a bus to get there. There would have to be satellite trial sites in places like Yarumal—another hurdle, another expense. Despite the fundraising success of Banner, which now had $15 million from donations and $15 million from the National Institutes of Health, the trial was projected to cost more than twice the original estimate of $50 million. Genentech would have to pay for most of it.

Ho began flying to Medellín every few months. Nothing got done in Colombia without an in-person meeting, and Lopera "was absolutely instrumental in any decisions that were made in Colombia, the crux of all decisions," she recalled. Lopera also proved a tough negotiator with both Genentech and Banner. If his registry was to reach its target of five thousand people, he told them, he needed more money; they responded with $3 million for the search. He balked at the language of the first contract presented to him, which reminded him too much of Novartis, but Ho traveled to Medellín to hash out their differences, and Kosik went with her to help. Genentech, mindful of Ho's security, hired a soldier to stalk the perimeter of Lopera's finca while the scientists tried to reach an accord.

For four years, until she left Genentech for another company, Ho kept up her travels to Colombia. She remembered those years as among the most exciting of her career. "It was so new," she said. "We were doing things that no one had imagined could be done." The other investigators on the trial described a similar feeling—of being part of a historic effort, of having incongruent forces come together as though propelled by something greater.

Perhaps the biggest accomplishment of all was Lopera's registry, which came to log some six thousand living members of the E280A families, a thousand more than were predicted to exist. Most had been recruited through the efforts of people like Ledy Piedrahita, though some had come in by way of a television and radio outreach campaign to families with cases of early dementia. The investigators also published news of the trial in *Q'hubo*, the trashy but useful Medellín tabloid that chronicled the comunas, hoping to find families unaware of their links to the larger clan and puzzled when dementia struck.

Of the six thousand people located, the investigators sought out three hundred suitable for the trial and willing to enroll: two hundred carriers and one hundred noncarriers. But they could not just hand them papers to sign. The E280A family members counted nine years of education on average. It was more than they'd had in the 1990s, but even still, few in the cohort had gotten beyond high school. Lopera began gathering groups in auditoriums. He began explaining the trial with a metaphor. Amyloid plaques, he told the families, were like garbage. When enough garbage builds up in the brain, you become sick. What did crenezumab do? Clean up the garbage. Lopera's team followed up with an exhausting campaign of slide presentations, question-and-answer seminars, and illustrated booklets. Only after all that did they present the families with papers to sign.

Participants in the trial would be compensated for every visit—a cash sum that was calculated not to be unduly enticing, as this was frowned upon in clinical trials. For a minority of participants, the money wouldn't matter; they were professionals forced to take time off from demanding jobs, or farmers for whom leaving behind animals for a day required onerous arrangements. For the rest, it would. A payment could mean the difference between having just enough for food and rent and having a treat—something pretty for the house, a toy for a kid, an evening out.

Whether the trial was ahead of its time was something the Banner scientists worried about. As did Ken Kosik, who feared that the science wasn't there yet, that the amyloid hypothesis didn't explain everything, and that a deeper understanding of the biology of Alzheimer's disease continued to elude them even as they took this costly, bold, risky step. "But then you land in Colombia," Kosik explained years later, when the trial was up and running. "And you see the plight of the families." Who wouldn't want to try anything that might help?

THE TRIAL NOW had a name: API Colombia. API stood for Alzheimer's Prevention Initiative, an entity created by Banner and its partners for its clinical trials. Months before the first shots of crenezumab were injected into the arm of the first patient, a group of Genentech executives visited Colombia, accompanied by the Banner scientists.

The drug company hired a military escort for their tour of the hospitals and clinics that would form the network of trial sites. Soldiers on motorcycles rode in front of and behind the caravan of armored cars as Lopera, Tariot, and Ho traveled with the Genentech team. Eric Reiman, perhaps dismayed by all the talk of kidnapping and Lopera's jokes about the FARC—you should always carry toothpaste, Lopera advised, because the guerillas never had any—stayed back at the hotel in Medellín.

The group spent no more than a few hours in the hill towns, not even daring to lunch in any of their restaurants. But they did make a brief stop at the church in Angostura, where in a room to the right of the altar, Padre Marianito's mummified remains lay enshrined. In previous decades, devotees had snipped off chunks of Marianito's appendages in the hope of procuring miracles, leaving him without ears or fingers. His restored body, with the addition of prosthetic hands and a lifelike mask, was unveiled in 2000, the year Pope John Paul II beatified Marianito and put him on track to sainthood.

You couldn't touch Marianito anymore or take a sample of his flesh, but in a receptacle at the iron bars separating him from the faithful, you could drop a petition. Into it the scientists dropped theirs, on a piece of paper folded up with some cash. They asked Marianito for the success of crenezumab, even if it took a miracle.

TEN

When I arrived in late 2017, the API Colombia trial had been under way for four years. The investigators never got the three hundred participants they had hoped for, and recruitment ended with just 252 people enrolled. Fortunately, fewer than expected had dropped out.

The CBS program *60 Minutes* had recently come to Medellín to meet the families and show the world what it meant to be a living laboratory for a disease. It was an optimistic moment in the public eye, though among the researchers themselves, worry was starting to creep in. Studies of crenezumab in patients with sporadic, or late-onset, Alzheimer's disease were showing that higher doses could be safely used, and the investigators feared that the Colombians were getting too little of the drug.

The investigators now wanted to switch the API participants to the high-dose infusions and extend the trial by several more years. They had no time to waste. The Alzheimer's research community had committed itself, with great fanfare, to having effective drugs by 2025. The problem was, 2025 was not so far away; in drug research years, it was practically tomorrow. A trial of a different anti-amyloid antibody, solanezumab, had recently failed in older people in the very first stages of the disease. Researchers knew that amyloid-beta went through molecular changes before becoming plaques and that these changes might be important; now they surmised that solanezumab had been aimed at the wrong "phase," or molecular form, of amyloid. Or perhaps, they said, echoing past drug failures, patients were recruited too late in the disease process to be helped. It was also possible that solanezumab had failed simply because the dose was too low, a predicament the API team was eager to avoid with crenezumab.

Management of the crenezumab program had recently passed to Roche, the parent company of Genentech. Things felt more corporate now. Without advocates like Carole Ho on the company side—people with a personal stake in the trial's successful completion—Roche could pull the plug at any moment, a scenario that neither Lopera nor the Banner scientists wanted to see.

In the first weeks of 2018, they, along with a group of very blond Roche employees who'd flown in from Switzerland, spent days holed up in a luxury Medellín hotel to work out the logistics of upping the dose and building a new center of operations at a university outpatient clinic a short walk from Neurociencias. In the new trial center, patients would recline in special chairs for hours with IV lines in their arms. It was a scenario Lopera had initially rejected, but the investigators felt they had no choice. "This is a moonshot—you don't turn the capsule around halfway," Pierre Tariot told me over coffee in the hotel lobby.

Tariot said that while it had been some time since he'd met with the E280A families—trial rules now prohibited sponsors from interacting with participants—he continued to regard them as sacred. It was a word they used a lot at Banner to describe the families, he said: *sacred*. Their pictures still hung in Banner's conference rooms, as a reminder that the researchers had to do everything in their power to get the trial right. "The worst thing we could do is hurt or betray them," Tariot said—including by keeping them on a low dose that the scientists knew, by now, had no chance of success.

LOPERA MADE TIME for me whenever he could. He invited me to weekend lunches at his hilltop finca, a modern estate that he'd named Monte Delphos after his youthful experiences in Greece. He and his wife, Claramónika, maintained their condo in downtown Medellín, but Lopera was most at ease on the finca, with its orchid trellises and fruit trees that he worked on in the mornings, and a pool where he swam laps. Half a dozen dogs followed Lopera around everywhere, and in the evenings he tossed grain to huge flocks of doves that waited for him in the trees, dependent on his largesse.

Though Lopera was not a practicing Catholic, and had only one child, a daughter, he was in many ways a traditional paisa. He ate his lunches at home, favoring meat and rice and beans. His social life revolved around his family, and his siblings and their children were his frequent guests. Every so often he reunited with his one hundred first cousins, who wore color-coded T-shirts to their gatherings to indicate which branches of the clan they belonged to.

His mother was now in the early stages of dementia. Doña Blanca was eighty-seven, and she spent her days filling coloring books or crocheting, but her designs had devolved into aimless worms of thread. Doña Blanca's illness was not caused by E280A and was nothing out of the ordinary given her age, but it touched Lopera's home life with some of the same sadness that the E280A families knew so well.

Often Lopera seemed amazed at his luck in having found such a rich vein of study; he was genuinely thrilled by the constant twists and turns of the research his group was conducting. Though at this point he spent only a few hours each week seeing patients, he still considered himself a doctor first—a value he shared with his American friend and colleague Ken Kosik. Lopera said he understood why Kosik seldom passed up a trip to Colombia: as a physician, he had to see the patients with his own eyes, talk to them, get a feel for them.

As much enthusiasm as he brought to the clinical trial, and as hard as he'd worked to make it happen, Lopera maintained a measured, agnostic attitude. His job, he explained, was to conduct the trial to the highest standards. If crenezumab worked, it worked; if it didn't, it didn't. As fond as he was of the families, he knew enough not to make promises.

AT THE UNIVERSITY of Antioquia's research complex, the Sede de Investigación Universitaria, or SIU, the researchers invariably referred to Lopera as *el doctor*, as though there were only one. Beneath their shared deference to the boss, I soon learned, they were given to shifting alliances, throwing in their lots with Lopera's various outside collaborators. These included Ken Kosik and some of Lopera's former students who had gone on to make their scientific bones in Europe and the United States, all while working extensively on data from the Colombian families. While everyone professed to get along, and their names appeared together on papers, the researchers jockeyed behind the scenes for prominence, data, funds, and Lopera's favor.

When there were no investigators to chat with in the SIU cafeteria, I took walks. There are things that jump out at you when you first arrive in a new place, only to fade from notice over time, and in the years that followed it became easy to forget that Neurociencias sat along a boulevard whose livelihood was death. At the south end was the stately Jesus Nazareno church, which performed funerals all day long. The church sat amid a long row of mortuaries, including one with a fleet of classic hearses—old Cadillacs and Buicks and Packards—to suit every taste. Everywhere in the district were young men and women who worked for the funeral homes, looking like flight attendants in their matching uniforms. Their job was to provide a somber yet attractive escort to the deceased on their final journeys.

This corridor of death continued north past the University of Antioquia medical school, where a mural of the slain doctor Héctor Abad stared benevolently onto the sidewalk, and the sprawling, verdant San Vicente hospital campus. Beyond that was a string of casket makers and florists and engravers who finished the stone tablets marking tombs in the city's mausoleums. Back in the day when the young Pablo Escobar was stealing and reselling these tablets, they were simple rectangles of polished stone; now they were made of a composite material that could be printed with color images of the deceased and their favorite things: the Virgen del Carmen, a motorcycle, the Atlético Nacional soccer team. The mile-long corridor of death took a turn past the hospital and terminated at an old cemetery called San Pedro, where mausoleum rows in concentric circles held thousands of crypts. Medellín's industrialist families occupied the coveted space at its center, shaded by cypress trees and statuary, while the poor rested in the surrounding rings, even in death held at arm's length from the seat of money and power.

THAT DECEMBER, I MET the E280A families for the first time. Since the onset of the crenezumab trial, the Neurociencias team had hosted an annual Christmas party. This time, the party was compulsory: all participants and their trial partners—designated family members, usually a spouse, who accompanied the participants on study visits— were required to attend. It was an unusual condition, but Lopera had news to deliver. He was announcing a radical change in the trial design: a major increase in the dosage of crenezumab and a conversion from simple injections to intravenous infusions. In order for the trial to produce a meaningful result, most participants would have to agree to the change. They would also have to remain in the trial through 2022, years later than originally projected. It was a big ask. A participant who'd joined in 2013, in the first wave of enrollment, would by then have given nine years of his or her life to crenezumab—without even knowing if he or she carried the mutation.

The staff transformed the courtyard of the SIU, erecting tents, setting up face-painting stations and a trampoline for the kids. They printed pamphlets to accompany Lopera's speech, complete with a friendly caricature of *el doctor* wearing a lab coat and gloves. When the families began arriving at 9:00 a.m., Lopera stood to greet them, hugging mothers and shaking hands solicitously with little boys, treating them like men. People surrounded Lopera, wanting a picture with him.

A few of the guests looked like farmers, but most were from in and around Medellín. They dressed the way people here did on special occasions: to the nines. Women wore platform heels, skirts slit to the thigh, and busty tops with spangles, all to sit through hours of lectures on amyloid-beta, followed by lunch and some music.

The podium in the SIU's auditorium was etched with an odd phrase: *Ciencia con Alma*, or Science with Soul. I wasn't sure what that was supposed to mean—did science have a soul?—but it did seem appropriate on that morning of kisses and hugs, with Francisco Piedrahita welcoming cousin after cousin. In my years as a science reporter, I'd never seen anything like it. My experience of clinical trials involved sitting in frigid hotel ballrooms as investigators presented data in tight twenty-minute sessions. Few participants in those trials would have met an investigator or any of their fellow participants, much less attend a party with them. Here I was watching a whole community converge to spend time with a trial's principal investigator and his colleagues—some of whom were, literally, family.

I settled in between Sonia Moreno, a longtime neuropsychologist with Lopera's group, and a farmer named Aníbal, from Canoas, as Lopera addressed the packed auditorium. In the years that followed, I would never admire Lopera more than when he was in front of the families. Even when he had bad news to deliver, as he did that day, he seemed to relish being in their presence, tailoring scientific concepts for their consumption, cracking jokes to lighten the mood. Today he delivered the bad news first, gently but frankly: The trial would be prolonged. For everyone. Until 2022.

This provoked predictable murmurs of displeasure, and he quickly pivoted to good news. To date, the investigators had seen no sign of serious side effects associated with crenezumab. The trial had received loads of international attention, and the *60 Minutes* episode, in which some of the family members had taken part, had been seen by millions. Lopera took the opportunity to explain, once again, the amyloid hypothesis of Alzheimer's disease and remind his audience that E280A carriers had brains that were ticking time bombs, already filling with "garbage," his pet term for plaques. He and his colleagues had seen new data showing that the crenezumab injections were not helping. "Now we know that the dose has to be higher," he said.

I looked to my left and saw Aníbal sleeping.

Was this a sales pitch? It did have that feeling. Lopera emphasized that while the trial's sponsors, Banner and Roche, wanted to make the dose change obligatory for everyone, he had refused that request. The infusions were monthly, instead of biweekly, but they would take several hours to complete. Those who wanted to keep receiving the low-dose injections could.

After his presentation Lopera took questions and comments, surprisingly few of which concerned the dose change. When will we finally learn our genetic status? someone asked. After 2022, Lopera said. This caused another round of murmurs.

One woman asked about the failure of solanezumab. The experimental antibodies targeted different phases of amyloid as it aggregated, Lopera explained, and none of the experimental drugs managed to attack all of its forms.

Sonia Moreno whispered to me that the woman's question was unusual. Few of the trial participants even referred to the study drug by its name, she said; instead of saying crenezumab, they just called it "the vaccine."

Another woman raised her hand to tell Lopera she felt she was in good hands. If he recommended the higher dose of crenezumab, then she would convert to the higher dose. "That's more typical," Moreno said.

With that, the science part was over. Francisco Piedrahita presented Lopera and Lucía Madrigal, who had rejoined Neurociencias after many years in Spain, with portraits by a local artist depicting each in the hills as a young investigator. He thanked them on behalf of the families of which he was part, "for this study that started in the mountains of Antioquia under the sun and rain."

The day ended with concerts in the auditorium, the families singing along to Mexican *rancheras* and Antioquian country songs as face-painted kids darted about the room. Despite the jarring news, the event was a success.

Late that afternoon, after everyone left with their takeaway boxes of Christmas pastries, Lopera gave me a ride home. I noticed that he had a copy of *Q'hubo*, the local crime-and-mayhem tabloid, in the car. He said he read it to see who was killing whom in the neighborhood near his finca, just to be in the loop. He asked what I thought of his presentation: Did people seem to follow it? He never quite knew how far to go with the science, he said, and probably erred on the side of too much information. On one end of the cohort was the lady who'd brought up the solanezumab results. On the other end there were the handful of participants who still believed that the disease was a result of witchcraft, even as they came in dutifully for their shots.

When I mentioned the question about genetic disclosure, he acknowledged that the families asked him about it every year, and that the pressures to end his ban were mounting. The API Colombia protocol stipulated that genetic results be offered to all 252 participants at the trial's end, but Lopera had just tacked on three years to that date. As for the thousands of family members not taking part in API, but who lent themselves to other studies, it was anyone's guess when or whether disclosure would ever become a reality. There were labs in Colombia that could, and did, test for presenilin mutations including E280A. But the tests were costly, and most participants in Lopera's studies, reliant on him for news and advice, weren't even aware they existed. The relationship the families had with the researchers appeared to be based on a default sort of trust. As Francisco Piedrahita pointed out, only Lopera and his team had ever bothered to go out into the hills and find them.

Part 2

Las Familias

ELEVEN

The Neurociencias brain bank was run by a physician named Andrés Villegas, a tall, somber man in his fifties, and staffed by medical students, graduate students, and residents. The research group had now collected more than three hundred brains from the families. They communicated through WhatsApp messages, as most people did in Colombia, and had to be ready to mobilize at all hours. Brains were supposed to be harvested and preserved within six hours of death, the sooner the better, and everyone left their phones by their beds, so as never to miss word of a donation.

Days after I joined that WhatsApp group, in February 2018, I got news of one. It was two in the morning when I saw Villegas's message: **Tenemos donación.** Then: **We need to move.**

The patient had died an hour earlier. By 4:00 a.m., the medical students were gathered in a dark hallway of the SIU, with their black rolling trolley full of equipment and a shallow white bucket. Their professor arrived unshaven but formally dressed, as Villegas always was. Years before, when he had spent time at Ken Kosik's lab at Harvard, he'd shown up every day in a shirt and tie.

The body awaited the team at a funeral home just a block away. Having Neurociencias so close to the corridor of death was convenient when you needed to extract a brain, examine it, section and photograph it, label all of the slices, and get one half of it into the tissue freezer and the other half into preservative within a six-hour window. Villegas walked ahead of his students, making long strides across the empty boulevard, papers clutched in his arm. At the funeral home, men in black suits offered him coffee. They knew him well.

This mortuary was Medellín's biggest—a huge, white, windowless chain of buildings that took up half a city block. Its dozen steel tables were in use even in the wee hours, as bodies continued arriving naked, in clear plastic bags, and going out dressed, in particleboard caskets. Villegas was supposed to conduct his brain autopsies in the hospital's morgue, but the funeral home's facilities were better, he told me. It all worked out so that the wakes could be held soon after the brain was extracted, with less trauma to the families.

When the students arrived with their bucket and trolley, Villegas sent them inside to start taking notes on the body, then went to wait on a sofa for the family to arrive. No brain could be taken without two family members signing off, but even if they'd long ago agreed to the idea, wrangling those signatures in that moment could be fraught. One needed to be focused, gentle, and patient, even as the clock was ticking.

Who was the deceased? I asked Villegas.

An E280A case, he said—and an important one, because the woman's problems had begun early, in her thirties. That was all he knew at the moment.

I returned to the preparation room, where the woman's body was curled in a fetal position on a steel table, not quite cold. She was wearing a clean diaper, with gauze wrapped around her head and chin. Delicate and fair, with wavy graying hair, she looked to be in her sixties, though I knew she was probably younger. Her skin was largely unblemished, without the severe bedsores most patients with advanced Alzheimer's developed. She was not emaciated, though her legs were stiffly bowed, indicating she'd been in bed a long time.

The students peeled the gauze from her face, making notes on her full eyebrows and eyelashes, her pale green eyes, her slightly aquiline nose. They grabbed at her tiny wrists and ankles, palpated each of her limbs. Her body was in good condition, they noted, which meant she'd been well cared for. The students drew samples of blood and cerebrospinal fluid, with the more experienced ones guiding the others, while morticians in scrubs worked quietly at their own tables, irrigating stomachs with metal hoses or stuffing cotton into noses as the radio played in the background.

We were joined by a tall, thick-set man in green scrubs, whom I'd never seen around Neurociencias but whom everyone seemed to know. His name was Esteban Muñeton—everyone just called him Muñeton—and he had a dark swagger about him that was more undertaker than scientist. At 6:00 a.m., the Colombian national anthem began playing on the radio. It was still playing when Villegas came flying through the doors: "You can begin," he said.

Muñeton took over and the students stepped back. He parted a spot on the back of the woman's head with a scalpel, holding the rest of her hair in his other hand as he assessed the part. Then he made a big, confident cut from ear to ear, and quickly peeled the scalp and much of the forehead skin forward, letting it fall over the face like a loose sock. The exposed cranium was oval, white, and bloodless. Muñeton turned on a circular saw and began to cut bone.

When the cut was complete Muñeton pried the skull open with the scalpel, making a loud cracking noise. He removed the top half of the skull completely. The brain went into the bucket, and the students took the tissue samples they needed for other studies: a cut of skin from the wrist, a snip of mucous membrane from the mouth.

Muñeton stuffed cotton wool into the woman's hollow skull, pulled her face and hair smoothly back over it, then threaded a needle with a white filament. He sealed the hair and scalp from behind, with large baseball stitches, and she looked no worse than when she'd come in. The morticians took over, starting to rinse and shampoo, as the students departed with their black trunk and white bucket, headed for the lab.

I caught a glimpse of the paperwork Villegas carried. The woman's name was Doralba López Pineda. *Doralba*, one of those melodious country names that evoked another time. She lived in Medellín, but had been born in Angostura in 1968. This made her forty-nine or just fifty.

It was 6:40 a.m. and outside the mortuary it was full daylight. I followed Villegas back to the dissecting lab at Neurociencias, as Muñeton walked the other way, toward the San Vicente hospital. I asked Villegas about the mysterious Muñeton. He was a prosector, Villegas said—an expert in dissecting bodies. The profession had been phased out in the United States and Europe, where autopsies were rarer than they once were. Muñeton worked in the hospital morgue, and Villegas just borrowed his services whenever a donation came in. "He's really good," Villegas said. "He does it perfectly every time."

IN THE DISSECTION LAB, Doralba López Pineda's brain was assigned its case number. It sat in its bucket while the students printed off a series of labels, and Villegas put on his mask and apron and gloves.

A third of the brains in the bank were E280A cases; the rest were from people with late-onset Alzheimer's or other neurodegenerative diseases, such as Huntington's or Parkinson's. A small number came from donors without any brain diseases.

Villegas removed a long, blunt-tipped knife from its butcher paper wrapping, and started to sharpen it under running water. This was a salmon knife, he explained. It worked well cutting soft tissue, and a fresh brain is even softer than salmon. He removed this brain from its bucket, and laid the meninges, or outer membranes, flat like a towel. He then started turning the brain over and over in his hands.

"This brain is really clean," Villegas declared. It was free of the arterial plaque you almost always see in Colombian brains by middle age, he said, the result of a national diet heavy in fried foods and pork. There was no evidence of stroke or other disease. This was one of the reasons these E280A brains were potentially valuable for science: by lacking the usual age-related damage that can confound studies in older people, they offered researchers a clearer view of how Alzheimer's acts on the brain.

Villegas started snipping parts and placing them into labeled vials. He took a sample from the olfactory bulb, a structure affected early in the disease course, affecting patients' sense of smell. He took other bits for cultures of astrocytes, spindly star-shaped cells that were of increasing interest in Alzheimer's research, as they appeared to play a vital—but still poorly understood—role in the inflammatory response of the brain. The field was now focusing not just on amyloid plaques and tau tangles, but on pathways and processes that might interact with those lesions or damage the brain independently.

Years of neuronal death had left this brain weighing 941 grams—about a quarter less than it should have. Villegas sliced it by hand, the old-fashioned way. There were other methods to ensure uniform sections, he said, but he had yet to find a device that worked as well as a salmon knife and his own eyes.

The students got their camera going, photographing each section on a blue plastic tray. By the time they finished, at 8:30 a.m., the left half of Doralba López Pineda's brain was reduced to nine uniform slices, each the thickness of a piece of sandwich bread, labeled in numbered bags, and beginning to freeze. The right half remained in the white bucket, now bathed in preservative. Villegas, looking downright haggard, took off his apron, washed his hands, and went to teach a class.

TWO WEEKS LATER, I met Villegas again on the university campus. He was about to spend several months in Hamburg with his collaborator Diego Sepulveda-Falla, a former Neurociencias physician who now ran a lab in Germany. Together they would perform immunohistochemistry studies on some of the brain bank's most vexing diagnoses. Villegas's students were helping him place slices of preserved brain into paraffin cassettes and delivering the samples to the airport.

While the Neurociencias brain bank lacked the equipment to carry out sophisticated pathology studies, Villegas said, it boasted more clinical data on its cases than most brain banks. This was thanks to Lopera's natural history studies that had begun years before. Look at Mrs. López Pineda, Villegas said. She had come into the clinic annually since 2001.

It was midday, and Villegas was just then waiting for the daughters of Doralba López Pineda to arrive. He would share with them the results of their mother's brain autopsy, which he carried in a manila folder.

When they entered Villegas's consulting room, I was struck by how much the two women resembled their mother. The younger daughter, Daniela, was twenty-five years old and apple cheeked, wearing a flannel shirt and a big silver cross around her neck. The elder, whom Daniela called La Flaca, or sometimes just Flaca, was thirty-one and had the slender nose and fragile build of Doralba. Flaca's young son dozed on her lap as they sat in the consulting room, where Villegas faced them from behind a desk.

Daniela carried a notebook and did all the talking. Villegas did not have a chance to share the first result before she spoke.

"I'm scared," she told him. "Our mom progressed so fast. I don't know if we're gonna get sick." Flaca was starting to have severe headaches, just like their mother had in the beginning, Daniela said.

Daniela wanted some assurance that their mother's brain was being put to good use, use "that could result in a cure." She didn't want future generations to suffer like her mother had. "I'm here because I want some help so that we can take care of ourselves, to prevent this, to avoid it," she said.

Villegas started in with what I took to be a standard speech. His grandfather, he told the sisters, had given him valuable advice: "Prepare as though you will live a thousand years, yet live as though you'll die any day."

Daniela, unmoved, asked him point blank whether she could learn whether she carried the mutation.

Not here, he told her. Neurociencias didn't have a genetic counseling program.

In that case, she explained to Villegas, the minute she detected the first symptoms of the disease, she would inject herself in the neck with an empty syringe and cause a fatal embolism. She already owned the syringe, she said.

Villegas nodded calmly. "A lot of people in these families say they will kill themselves if they get sick," he countered, but in the end they don't. "Why do we live?" he asked her. "What do we live for?"

Daniela thought about that. "Because we assume we have a future," she said. "A long-term plan, a profession, a house, a car, some travel." She wouldn't kill herself right away, she said, while these things were still within her reach.

Perhaps cued by the cross around her neck, Villegas tried a religious approach. "As Catholics," he said, "we're not obliged to be miserable—Jesus wanted us to enjoy ourselves as well."

Daniela wasn't having it. "For me to go through this, and to put my family through this," she said, shaking her head. "Maybe it's cowardly, I don't know. But it's my solution." If her mother had killed herself early in her disease, she said, it would have been fine. "The way they end up is just"—she could not finish her sentence, tearing up as she talked.

Villegas told her that his brother had killed himself, and that the suicide had left a terrible scar on his family. I started to wonder whether his pro-life discourse reflected his own deeply held beliefs, or whether it was what resonated best with the thousands of pious Catholics who'd streamed through Neurociencias all these years.

I'd also seen Villegas express a lot of compassion toward the families. He even barred the brain bank staff from chatting in the presence of a dead body, as a gesture of respect. Villegas had his doubts about the crenezumab trial, but he badly wanted the drug to work, because the families needed hope.

Right now he was an hour late meeting his students, who were waiting for him in the lab to start cutting more samples for Germany. Yet he did not glance at his phone or watch as he worked methodically through the list of autopsy findings. Daniela asked about the tissue samples taken from her mother's mouth and wrist: What cells were being cultured? What experiments were being conducted? And was all this in the service of finding a cure? She scrawled the answers diligently in her notebook as Villegas walked her through a litany of terms: *Astrocytes. Glial cells.*

And then, as though finally recognizing the kind of young woman he was dealing with, Villegas addressed Daniela frankly. Since she was already approaching her future as though she knew she would become ill, it might indeed make sense for her to get tested, he told her. "Otherwise, you're living your life around an assumption."

If she turned out not to have the mutation, he asked her, would her plans change?

"Obviously," she said. She would like to be a mother, she told him; it was the thought of another generation suffering that stopped her. "I don't want any part of prolonging this," she said.

Villegas nodded sympathetically as he rose from his desk.

"This disease is really fucked up," he said.

DORALBA LÓPEZ PINEDA had been dead a month when Lopera shared a copy of her file with me. On her first visit, in 2001, a thirty-three-year-old Doralba had told the doctors that she was a housewife with two daughters and a sixth-grade education. Her mother had died of Alzheimer's disease in 1992, at age fifty-four.

She has no memory complaints, Lopera wrote of his patient. *Nervousness: from childhood. Sometimes she cannot sleep for fear that someone is going to hurt her. She sees a little owl approaching her. She does not know if she is dreaming it or seeing it.*

Lopera was just then starting to follow the healthy children and grandchildren of E280A cases. He tested subjects for the mutation, not sharing their results with them. Every year or two they came in for a checkup, and they were invited to participate in different neuropsychological studies, which often took the form of group activities alongside their own relatives. Some were tedious games in which you had to count images or objects. But participants were paid a small fee to come in, and people like Doralba needed every cent they could get.

In 2005, Doralba appeared at her checkup depressed, which Lopera, in his notes, ascribed to an "external event (husband in jail)." She stumbled solving any problem that had to do with numbers.

At forty-four, she reported severe headaches. Her memory faltered. She would leave to go shopping and have to turn around and come home. She began to lose money, and no longer left her house except to go to church.

At forty-six, Doralba López Pineda was determined to have full-blown dementia. A team from Neurociencias visited her at home and found her in bad shape. "She is always swearing and screaming in a hallucinatory state as if she were talking to someone," they wrote in their report. "She also sometimes thinks that someone is grabbing her arm. She likes to bathe all day and washes clothes for hours and hours. She likes to be naked. She went out naked last night in the neighborhood and her daughters lost her. She hits her daughters. She gets angry for no reason and starts screaming, then switches to laughing."

The report described a common pattern, an almost choreographed progression in the histories of the E280A patients. The age at which Doralba became impaired, forty-four, was standard. People turned on their own family members, often lashing out violently. Running around naked was not unusual either.

But a little owl, flying toward her at night? That was strange. Doralba was a young woman when she reported that, years away from her illness. I wondered what it could have meant.

TWELVE

I'd asked both of Doralba's girls, Daniela and La Flaca, to meet me somewhere away from Neurociencias and tell me about their mother. Only Daniela, the outspoken one, agreed.

We waited for each other a full hour before realizing that we were on opposite sides of a huge train station, the last stop on a metro line terminating at the base of a giant hillside district known as Comuna 13. From the station, the comuna—a city within a city, home to more than 150,000 people—looked like an explosion of brick radiating up into the mountains.

When we finally connected, Daniela was wearing the same big silver cross as she had that day in Neurociencias, with heavy makeup and her blond hair in a braid; her nails were painted a metallic lime. Though her WhatsApp icon was a drawing of a woman with her middle finger raised to the world, she could not have been friendlier. We drank iced coffee in the street by the station as the metro's cable cars sailed by overhead.

I asked Daniela what she'd thought of her conversation with Andrés Villegas. It was the kind of thing she'd been hearing her whole life, she said. Daniela and her sister, along with their first cousins, had grown up as subjects of Neurociencias, evaluated every few years.

Daniela was eight the first time she accompanied her mother to see Lopera. She'd since participated in different studies whose details she knew little about, and she never learned the results of any of them, which annoyed her. She played along, despite her misgivings, because she felt obligated to help the researchers, and because she hoped for the chance to be part of a drug trial someday. She and her sister had been too young to take part in API Colombia. La Flaca had wanted to enroll, but had missed the age cutoff by a year.

I felt the need to confess, before we got any further, that I'd witnessed her mother's autopsy. Daniela took that well, and she replied with a confession of her own: on the morning of the autopsy, she'd asked Andrés Villegas to preserve her mother's eyes for her.

"In a jar or something?" I asked.

"Uh-huh," she said.

"And what did he say?"

"He told me that was very strange," she said, starting to laugh. "But my mom was beautiful, she really was. Her eyes were a dream."

One of her uncles had the same eyes, she told me. He was sick now. Two of her mother's sisters were also gravely ill. The youngest of Doralba's siblings, who was forty-four, appeared to be in the first stages of the illness. Another sister and a brother were already dead. Three other siblings were fine. I counted that up: of ten siblings in her mother's generation, seven were sick or dead.

In two months Daniela would turn twenty-six. She spoke well and she spoke fast, almost without pause. At first she had to shout over music playing in the tavern next to us as she told me her story, which was in large part her mother's.

IN THE LATE 1970s, Doralba López Pineda and her family migrated to Medellín from the town of Angostura after their father, Abel, died of a heart attack. Abel was in his forties and a tailor, a well-liked man about town. He left behind a wife and ten children.

The siblings and their mother set themselves up in a neighborhood called Pedregal, an area close to the bus terminal that was a landing point for many new arrivals from the hill towns. At first some worked as regional drug mules, shuttling packages by bus between the country and the city. Later they got jobs in factories, or as live-in maids, or doing construction.

Doralba, a skinny, rebellious blonde, never truly found her way. She was good-looking and vivacious and did as she pleased, working odd menial jobs. Doralba was still a teen when La Flaca was born, to a father no one knew. When she became pregnant with Daniela, some years later, her older brother Jaime kicked her out of the family home. Jaime had become the de facto head of the family and it was he, not their mother, who had the final say.

Doralba went to live with a friend who had daughters of her own. She gave birth to Daniela, and the women alternated working and providing childcare, though only Doralba was any good at either. The friend was drunk all the time, disappearing for days and returning crazed and soiled, lying in a heap outside the house. Worse, she had a boyfriend who tried to touch the older girls in their sleep. Doralba, lacking the means to move out, slept in the bed with the children, her head at their feet and her arms stretched protectively over their genitals. "This went on for years," Daniela said, until the boyfriend was murdered over something else.

The girls grew up skinny, dependent on people's pity and their own wits and Doralba's creativity in the kitchen. They ate soup donated from neighbors, cracked eggs they scavenged from crates behind a produce store, and Bienestarina, a government-issued nutritional powder that Doralba threw into milkshakes and desserts. Despite their poverty, Doralba dressed her daughters as carefully as dolls. "I was blond, very white, with pink lips," Daniela said. "And more than once someone said to my mother, 'Let me have her—you can't provide for her.' But she never gave me away, and I'm eternally grateful for that."

Doralba was selling arepas from a street cart when she met the man who would change her life. Orlando was tall, handsome, and a few years younger. He had come to the city recently from a remote coffee-growing town. Doralba's girls thought she seemed spellbound by Orlando, a reaction they'd never seen before in their mother. When Orlando learned of a new invasion under way in Comuna 13, on the other side of the city, he and Doralba jumped at the chance to make a home of their own.

Invasions are collective land grabs in which poor people stake claims on vacant land, taking advantage of their strength in numbers and the government's unwillingness to stop them. If officials don't act quickly to drive the newcomers out—often because an armed group is protecting them—their hastily built homes can one day become titled, legal properties. That was the gamble Doralba and Orlando took when they bought a steep plot from some *milicianos*, or militiamen, who didn't actually own it. The milicianos were local youth, aligned with the guerillas, who every so often would pull populist stunts like robbing a dairy truck and handing out the milk. They policed the plots, assuring that no one could edge out their designated occupants. Doralba and Orlando paid the milicianos the equivalent of nine hundred dollars for their plot, in installments.

Orlando, who had all the country ways and skills, knew how to build a house from nothing. At first the family scavenged old wood and palm fronds. They hung pots to collect the rainwater that poured in through the thatched palm roof, and stored food in plastic containers to keep it away from the river rats that crawled up from a creek. Their first toilet was a wooden box lined with a garbage bag. Orlando returned to the countryside for months to pick coffee, and Flaca left for long stretches to live with her better-situated uncle Jaime, leaving Doralba and Daniela to sleep huddled in a bed above a soggy earth floor.

In time they put up a metal roof, spread cement over part of the floor, and built a well in the kitchen that they used as a refrigerator, tying up their bags of milk and submerging them in its cool depths. Doralba hung curtains of every color on her wooden walls, even where there were no windows. One of the nice things about the comuna was how rural it felt. Since the 1980s, it had attracted migrants from the countryside. Men sat outside with machetes in holsters, grew plantains and coffee, and kept fighting cocks; they painted their rickety porches red and yellow, turning the shanties into little replicas of fincas. Kids flew homemade kites. In the mornings, Doralba listened to the country melodies of Radio Paisa as her pressure cooker hissed, cooking lunches she wrapped in banana leaves for the girls to take to school. She raised rabbits and chickens and grew herbs for medicine, a rue bush for good luck.

To Doralba, Orlando was a white knight, the man who built her and her girls their first and only home, who promised never to let them go hungry. To the girls, Orlando's arrival marked the transformation of their caring, protective mother into someone more compromised. Orlando was indulgent toward La Flaca, by then a pretty teenager who was a dead ringer for her mom. But Daniela was not yet ten, fierce and blunt, when she and Orlando began to clash. Orlando drank what was known in Medellín as chamberlain, a street cocktail made from rubbing alcohol, milk, and powdered drink mix. The drink brought out his sadistic tendencies. He charged at Daniela with a machete, snapped mousetraps on her fingers, and locked her out of the house, forcing her to sleep with relatives or friends. When she was in the fourth grade, Daniela published a short story in her school newspaper about a butterfly with no place to land: "One sad, melancholy and rainy afternoon, the golden butterfly flew around and around, without stopping," it began.

By the early 2000s, Comuna 13 was becoming increasingly dangerous. Police and gangs, sometimes working in tandem, targeted the militias. Families that had fled violence in the countryside now had to soothe children who were being kept awake by gun battles. Many found themselves accused of colluding with the militias and were forced to abandon their hard-won homes. Doralba and her family, though, were spared. They survived a takeover of the comuna by government forces in 2002, holing up in their house for days as helicopters shot at militants from the sky. Daniela created a homemade weapon out of sweat socks and rocks that she swung over her head whenever she had to walk outside. She started sleeping with a hammer under her bed.

As a young teen, Daniela began plotting her stepfather's demise. She grew furious with her mother for allowing his abuses. "I forgive her—I forgave her a long time ago," Daniela told me. "But there was a time when she was more a woman than a mother. Or she became more of a woman and forgot she was still a mother."

Daniela tried to kill herself by drinking bleach. She failed and was admitted to the Hospital Mental. Doralba came to a few therapy sessions, but seemed loath to acknowledge Orlando's role in bringing Daniela to despair. "I tried to understand that she was scared of him," Daniela said. She filed a formal complaint against her stepfather in family court, which her mother refused to support. She started entertaining thoughts of more drastic solutions.

The government offensive in the comuna had given new gangs a chance to fill the vacuum left by the militias. The gangs were allied with different factions of the Medellín mafia and with the right-wing paramilitaries active in the countryside. The *paracos*, as people called them, staked out turf, creating borders out of creeks and stairways and defending them with bullets. As with the militias they had replaced, residents went to them with their problems. La Flaca approached them with a request phrased in terms they would understand: her family needed some "social cleansing." The paracos agreed. But when they finally came for Orlando, he escaped through a window, not to be seen again for weeks.

Later that year police arrested Orlando on an old warrant. The whole time Orlando was in jail, Doralba smoked, fretted, lost weight, professed his innocence, and cried. The E280A families often reported to the Neurociencias clinicians that a traumatic event had propelled their loved one irrevocably toward dementia. The clinicians remembered certain types of events—a son or daughter running off with the guerillas, for example—having marked the start of more than one patient's decline. Daniela would come to think that Orlando's jailing was the beginning of her mother's disease, an emotional prologue to the assault on her brain.

Daniela withdrew into school, spending time with a teacher and a small group of kids who liked to study together. But even studiousness could not protect them; one boy in the group was murdered for crossing an invisible gang border. Daniela started smoking marijuana, and once suffered delusions that she was being chased by the paracos, nearly landing herself in the hospital again.

La Flaca met a young man who would later become her husband, and Daniela felt herself falling in love with a friend, a girl. When she was about to turn fifteen, and Doralba tried to plan her a quinceañera, Daniela was forced to tell her mother she had no boyfriend to show off at the party: she was a lesbian. "I knew it" was all Doralba said, and she went back to washing dishes. It wasn't the warmest response, but for that time and place it was about the best Daniela could expect. Doralba, passive as she was in the face of Orlando's abuses, loved her spirited daughter and seldom found cause to criticize her.

One day, without warning, Orlando walked into a modest Pentecostal church and walked out declaring himself saved. Doralba soon followed suit. The sect had been established in Colombia in the 1930s by missionaries from Canada, and the rules from its founding era remained in force: taboos against not just dancing and alcohol but earrings, makeup, and pants on women. Doralba quit smoking, traded her sexy jeans and baseball caps for long skirts and leather-covered Bibles, and snubbed her old friends. There were no more statuettes of Padre Marianito, no Christmas-season processions from house to house, shaking dried seedpods like maracas and singing traditional songs. Radio Paisa was banished in favor of Christian music. Doralba refused to attend the Catholic baptism of La Flaca's baby boy.

Doralba married Orlando in a prim ceremony in their church, wearing a gray jacket and flowered skirt, her hair ironed flat and her face bare. After ten years with him in decadence, she now joined him in piety. "She was with him body, mind, and soul," Daniela said. The couple's conversion, jarring though it was, brought a stability to Daniela's life that she had not before experienced. Orlando's chamberlain-fueled rampages ceased and his machete stayed lodged in the rafters; stepfather and stepdaughter arrived at a détente. Daniela, once again the only daughter living at home, joined the dance team at school and kept up her grades and her scholarships, washing pots in the afternoons at an arepa factory. She looked forward to university. The fact that she was gay bothered some of Doralba's church friends, but Doralba ignored them.

As time passed, however, Doralba failed to thrive under the new religious order. By her forties, her outsize personality was diminished and she seemed to be receding from life. She'd always loved children and for a while even ran an unofficial daycare. But she showed little interest in her own grandson. At the time, Flaca and Daniela blamed the Pentecostals for distancing their mother from them. Later, looking back, they blamed her disease, which they had been slow to see coming.

Doralba's mother had died of Alzheimer's in 1992. She had been seen by Lopera and the clinicians. But there was little sense among Doralba and her siblings, despite their frequent contacts with Neurociencias, of what they should expect for themselves and when. "My mother's generation wasn't well informed," Daniela said. "That was one of the biggest mistakes we made as a family."

Daniela enrolled at a polytechnic institute. Her life was moving in a hopeful direction when she noticed that Doralba was starting to repeat herself. "She would tell me weird stuff about my sister as a kid," Daniela said, as though Doralba's past had become more vivid to her than the present. Memories fifteen or twenty years old flooded her conversations. Doralba began cleaning the house and her body obsessively, showering several times a day. She started putting things where she knew she would have to pass to find them. At night, unable to sleep, she cried and sang church songs.

"She was scared when she first got sick, not knowing how far all this would go, where this would take her," Daniela said. Daniela's studies got placed on hold as she, her sister, and Orlando struggled to keep the increasingly disoriented Doralba safe. They installed a wire gate in the doorway to prevent her from wandering. Lucía Madrigal and the Neurociencias team visited, writing orders for diapers and medicines to help Doralba sleep, but a hospital bed—something the researchers often provided to very ill patients—never arrived. Daniela hurt her back lifting her mother over and over. To bathe Doralba, Daniela had to carry her over a dirt floor into a bathroom precariously anchored to the side of the house, then return her to her bed under a crude metal roof. The investigators noted that Daniela appeared depressed. At night she smoked joints to ease her back pain and to escape.

Throughout her ordeal, Daniela maintained a long-distance relationship with a graduate student named Ximena, who lived in Bogotá. When Ximena visited, she was shocked by the conditions in which Daniela cared for a mother with advanced Alzheimer's disease. The house, Daniela knew, was ugly. But this was the home Doralba and Daniela had built and made theirs. As Doralba's language shrank to a few short phrases, the house became one of her recurrent themes. "The house," she'd repeat to Daniela. "The curtains—yours." The curtains were Doralba's delight. She had always washed them every week and rotated them out, storing extras in a hamper. "You see?" Daniela taunted her stepfather. "She's saying she wants it to be mine." Daniela credited pressure from the church for keeping Orlando around to help care for her mother. Before his religious conversion, she was sure, Orlando would have fled for the hills, and though Daniela didn't like Orlando in any incarnation, his presence—the mere sight of him—buoyed Doralba's spirits, even after she'd lost all her speech.

Ximena became pregnant and gave birth to a baby girl, and Daniela divided her time between two cities, between one life that was starting and another that was ending. In Bogotá she cuddled and fed Ximena's child and loved her like her own. In Medellín her role was essentially the same. She applied creams to her mother's body every day, keeping her skin soft and unblemished, cleaning up Doralba's feces and changing her diapers. Doralba had become as another baby to Daniela. Her face lit up when she tasted something delicious; she beamed at Daniela's cooing.

Daniela and La Flaca refused to cut their mother's wavy hair above her shoulders, though it would have made things easier for them. Washing and drying it was a concession to Doralba's vanity, to a womanhood she'd worn proudly and which no one wanted to rob her of. When Daniela, Flaca, and Orlando finally had to put Doralba in a nursing home, and her body curled into the fetal shape it would retain for the last year of her life, they still refused to let anyone cut her hair.

Doralba died on a Thursday night. She'd been hospitalized with a bout of pneumonia and was receiving sustenance through a tube. Her daughters had come to see her the previous Sunday, and both thought she was improving. Doralba could no longer respond to queries by blinking, as she once had, but her eyes remained bright and responsive. She laughed at the sight of her girls. "Look how well she's doing," Daniela said to La Flaca, who agreed. There was never a moment when they felt Doralba wasn't aware of their presence.

The next day, though, Doralba did not open her eyes, nor did she in the days that followed. Her fever returned. On Thursday morning, she began to splutter phlegm. "I couldn't understand what was happening," Daniela said, "but I felt that day that my mother was going to die." That evening, when she was alone with her mother, Daniela massaged Doralba's limbs as Doralba's temperature continued to rise, and her breath made a rattling sound.

Daniela went out to find a nurse, who told her gently that there was nothing left to do. She returned to her mother's room.

"I felt her take her last breath," she said. "She looked pale and I couldn't hear that rattle anymore." Daniela tried to move Doralba's limbs, not wanting to believe her dead. But the nurse confirmed it.

"I called La Flaca," she said. "She picked up saying: 'What happened?' At that hour, she knew something had to be wrong. I said, 'Mom just died on us.'"

Daniela let the hospital wrap her mother's head in a bandage, to keep her jaw closed. Doralba's body was still very warm, from her fever. Daniela knew what came next: Doralba would be removed to the funeral home where Neurociencias would be waiting to harvest her brain. Before signing the papers to release Doralba's body, Daniela changed her mother's soiled diaper and cleaned her, then applied more cream to her skin. "Because to me, at that moment, it was as though she was still alive," she said.

Doralba's funeral was brief and spare. A group came from the church to sing and pray around her casket before her body was taken to the crematorium. Only half of her siblings made it downtown to see her. The rest were too sick, or had already died.

The youngest among the siblings, María Elena, behaved strangely at the funeral home. She was confused about the whereabouts of La Flaca's son, whose father had taken him outside. Even after she was told where he was, she asked for the child again, sounding distressed. When it was time to go home, she began walking in the wrong direction. María Elena was slender, well dressed, and bubbly, the kind of woman who always lit up a room, but now Daniela noticed a sunken, distant quality in her eyes that reminded her of her mother in the earliest stages of the disease.

Daniela couldn't bear to return to the house in Comuna 13. Orlando could have it, curtains and all. She didn't want to see him ever again, even if it meant forfeiting her memories, her photos from high school, her mother's smell. She went to stay with La Flaca and her family, who had an apartment by the train station. She picked up some work at an ad agency, but within weeks she would return to Bogotá and Ximena, to try to resume her relationship and find direction in her own life, which had been half lived for years. She didn't know what she would do, just that she had to do something.

"I can't be like my mom," she said. "I have to be different. I can't be a *gamina*," some girl off the street.

I asked her if she still wanted to be tested for the E280A mutation. She did. She just didn't have the slightest idea where or how to go about it.

Unlike her sister, La Flaca didn't want to know her genetic status, even though she was concerned enough about the disease to try to get into the crenezumab trial. She suffered migraines, like her mother had, and Daniela feared for her. Daniela encouraged Flaca to play more games, read more, stimulate her brain. Their aunt María Elena had been screened for the API Colombia trial and invited to enroll, but she had refused. She had no need for the money, she claimed, and she was too busy.

AS IT GREW DARK, Daniela and I migrated to a chess table in the park to get away from the noise of the street.

I asked her about her suicide plan, the exotic and dubious death by syringe she had described to Andrés Villegas. Was she serious? Yes, she said. "Living with Alzheimer's isn't a life. It's humiliating. You look like a mummy. It erases your life, even your sense of smell." She was holding back tears as she spoke. Whatever she'd lived up to this point—all she had seen and suffered, all the ground she'd lost and would struggle to recover—was worth it, she explained, as long as she never got sick.

THIRTEEN

Ken Kosik was staying at the InterContinental hotel, the elegant fortress above the city where he had landed on his first, precarious visit to Medellín nearly thirty years earlier. He had about a month's worth of things to do and a week to do them in, as was typical for his trips to Colombia. He needed to make an inventory of the Neurociencias brain bank for a grant he hoped to get from the National Institutes of Health. He would review with Lopera and his team everything yet known about the other dementia-causing mutations he and his colleagues were discovering in their samples from Colombia. And he had the delicate task of chaperoning another Alzheimer's researcher, Randall Bateman, around Neurociencias, where they would meet one of what Lopera and colleagues were calling the "new" families—those whose genetic makeup included presenilin mutations other than E280A.

Lopera's recruitment campaign during the run-up to the crenezumab trial had uncovered hundreds of cases of young-onset dementia, but to the investigators' surprise, only a few new branches of the E280A clan. The rest of the cases were negative for the mutation. Kosik's team, analyzing the samples, had identified a handful of previously unknown presenilin mutations, but it remained to be seen whether they also occurred in large families, and how their diseases behaved.

To untangle all this would take years, but the group was working fast. That API Colombia had been successfully launched signaled to the international Alzheimer's research community that it was possible to conduct complex clinical trials in Colombia, and people like Bateman thought perhaps the new families might be candidates for one.

Bateman ran a research program at Washington University in St. Louis, Missouri, called the Dominantly Inherited Alzheimer Network, or DIAN. The network, which got under way at around the same time as the Banner Alzheimer's Institute, led important studies in familial Alzheimer's disease, putting together cohorts of people from different families worldwide with mutations on the presenilin and APP genes. The DIAN researchers had discovered, among other things, that the brain's ability to take up glucose became impaired a full decade before symptoms emerged, and that tau tangles appeared in tandem with the onset of the first cognitive symptoms. DIAN did not develop Alzheimer's drugs itself, but its clinical trials arm recruited mutation carriers to try different agents that pharmaceutical companies sought to test.

The Banner and DIAN scientists were too diplomatic to describe each other as rivals, but there was an element of competition all the same. Pierre Tariot and Eric Reiman were aware that Randall Bateman was poking around Colombia, and they didn't sound thrilled about it. Banner and Roche had built the entire infrastructure of the API Colombia trial, an expensive and ongoing investment. The unspoken agreement between them and Lopera seemed to be that the E280A family members, representing the largest early-onset Alzheimer's cohort in the world, were off limits to other trial groups. But the Colombian families discovered more recently, by Kosik and Neurociencias, were deemed suitable for DIAN's projects. It was just that Lopera and his colleagues still knew little about these families.

Physicians in the Colombian city of Cali had found one family with a mutation that caused symptoms starting in people's thirties. Lopera's group had recently discovered a larger family with a different mutation living in Montería, near the Caribbean coast. Finally, there was a clan that had first come to the attention of Neurociencias in 2002, whose members had an early-onset Alzheimer's mutation that behaved similarly to E280A. But it wasn't until 2015, after the crenezumab trial was under way, that the team started inviting them into the clinic to be studied. The family's unique mutation, labeled I416T for its switched amino acids and position on the presenilin gene, had yet to be reported in the scientific literature, and none of their brains had made it into the brain bank. Most of its members came from one town: Girardota, north of Medellín.

As soon as Bateman learned about them, from Kosik, he arranged to fly to Colombia. Kosik would help Bateman make his case to Lopera and the family—whom not even Lopera had formally met—that they should join his program and try new drugs, whatever they might be.

KOSIK HAD DESCRIBED BATEMAN as pastor-like, and indeed he fit the bill: a tall, pale Midwesterner with a tendency to speak in parables.

On the morning that Bateman arrived at the SIU, with colleagues from Washington University in tow, he eagerly took notes as Lopera described the Alzheimer's family they would meet later that day. The Girardota family, Lopera explained, had come to the researchers' attention through the case of a woman with dementia who had been suffering mood swings and memory loss since the age of forty. More than one hundred of her relatives had since been evaluated, he said.

Bateman, in turn, briefed the Neurociencias team on DIAN and its ongoing trials around the world. He included the intriguing detail that while the DIAN trials recruited both carriers and noncarriers from families with mutations, just like API Colombia, DIAN encouraged every participant to learn his or her genetic status through a counseling program, and let noncarriers drop out if they wished.

The Washington University team had apparently not been advised of Lopera's longtime policy against disclosing genetic information until after people became ill. At a lunch following the meeting, Bateman seemed incredulous to learn that the healthy people in Lopera's studies were not allowed to learn whether they carried a disease-causing mutation. DIAN's policy everywhere in the world was to give people the option of knowing, Bateman said, and the program included a generous budget for counseling. "Not here," Lopera replied, in English, in a tone more adamant than I'd ever heard him speak in. If the Girardota family members were allowed to learn their genetic status, it would be unfair to the E280A families, he argued. Bateman was calm but insistent. A different trial with a different family could have different rules, he said. Couldn't it?

No, Lopera said.

The exchange became tense, and Kosik, sitting next to Lopera, was uncharacteristically silent. He could be pugnacious with his American colleagues, as I would soon discover, but in Colombia he deferred to Lopera. By dessert, Bateman and Lopera had reached a temporary truce. They would look for a solution later. When the new family came in—they were scheduled to arrive in less than an hour—the issue of genetic disclosure would not be discussed, they decided.

The family's arrival was delayed, their walk from the metro to the SIU blocked by a march commemorating the thirtieth anniversary of the murders of Héctor Abad and his colleagues at the University of Antioquia. When the attendees finally trickled in, slowly filling the auditorium, it became apparent to me that the Girardota family was Black. I had met Black members of the E280A clan before, but they were rare. Nearly everyone in this room, by contrast, looked to have some African ancestry.

A middle-aged woman sat next to me chatting nervously. "This genetics is heavy—this is serious," she said. That there was a rare mutation in the family was news to her. Her sisters and a brother had died in their fifties of dementia, after being bedridden for years. But the doctors where they lived seemed to view their case as unremarkable. They'd called it "senile dementia," she said. In someone so young? I asked. Yes, she said. "We were so in the dark."

Before the woman could continue, the program began. For many in the room, this was their first time meeting the famous Dr. Lopera—they'd been seen until now by other clinicians in the group—and he did not disappoint. He was warm, dashing, funny. Some in the audience used their cell phones to record his familiar talk about garbage in the brain, the natural history of plaques and tangles. How at this point about two hundred presenilin mutations had been discovered worldwide, and that the Girardota family had one of them.

Bateman, when it was his turn, had a harder time connecting with the crowd. He spoke through an interpreter, talking at length about the DIAN program and its various arms. He explained how statin drugs, which lower cholesterol, had first been tried in people with rare genetic diseases that caused cholesterol to skyrocket, and that a parallel might exist with Alzheimer's.

After reciting an exhaustive series of caveats about exploratory studies, regulatory approval, and red tape, Bateman finally addressed his audience directly. "How many of you think you would want to participate?" he asked.

Hands shot up all over the room. "Everyone!" people yelled.

Bateman reminded them that a clinical trial would mean undergoing invasive procedures, including spinal taps; that it would require them to use birth control for years; and that they might have to fly to

Bogotá for imaging, as the radiotracers DIAN used were not available in Medellín. No one seemed dissuaded.

Someone asked if they could smoke pot. Before Bateman could answer, Lopera interrupted.

"Once a week," he said, to laughs. He'd dealt with this before. Anyone in the trial would have to give investigators a careful accounting of all the medicines they took, "including whatever the witches gave you."

Bateman resumed, somberly. He was keen to learn why this group wanted to participate in clinical trials, and what they thought the trials were for.

"For the kids," people answered. "For those to come."

"To break the chain."

Bateman did not respond to this; he may not have understood. He went on to answer his own question: "It's to see whether medicines work," he said, and returned to the subject of statins. When he finished his talk he was surrounded by people eager to volunteer. They were not there to save the rest of the world or to see if medicines worked. They were there to save their children.

THE LONG-HELD CONCEIT in Antioquia, one shared by some of its best-educated people, is that Black people are not original to the local population, but rather later arrivals. But an academic paper I found online, shortly after Bateman's visit, described Girardota and its outlying hamlets as home to a large population of Black montañeros whose African ancestors had arrived in the first half of the seventeenth century. One of their hilltop communities had long been of interest to historians and musicologists. Its families of professional musicians maintained a tradition centered on string and wind instruments. The same families also preserved a form of Spanish comic musical theater known as *sainete*. Sainete had long died out in Spain, but in the mountains above Girardota it was alive and well, performed exclusively by Black paisa troubadours.

When I had lunch with Kosik at the InterContinental hotel, talking over the faint rhythms of tennis balls being whacked, he was intrigued by the idea that the Girardota family might have a unique history, just as the E280A kindred did. He had been part of the team that visited the first patient in her home, back in 2002, thinking that the family had E280A. The fact that they appeared to have African ancestry hadn't struck him as important then, but it did now. His team had begun to look closely at ancestry with the new mutations, in part because other investigators had been finding that ancestry could alter a person's risk of developing dementia, or could affect its course. The APOE4 variant, for example, was emerging as a stronger risk factor in people of European descent than of African descent. But it was not a person's overall ancestral makeup that mattered so much as what was called local ancestry—the ancestry of the DNA surrounding the variant. A mutation that caused severe disease on a Europe-linked haplotype, or background, did not necessarily have the same effect on an African, Asian, or Indigenous haplotype. This was because different genes surrounding the mutation can change its expression, Kosik explained.

The ancestral backgrounds of the new mutations had cultural implications as well. Kosik and his colleagues had previously established that the E280A mutation sat on a haplotype linked to Europe, which seemed to confirm local beliefs about Antioquia being whiter than other regions of Colombia. In the early 2000s, this idea was promoted by a group of University of Antioquia geneticists who set out to show, in paper after paper, that Antioquia was a "genetic isolate," in effect an island of whiteness where rare European mutations thrived. But if the Girardota mutation was found to be linked to Indigenous or African ancestry, the idea that racial purity was at the root of this regional epidemic could be tossed right out the window.

In fact, to Kosik, the opposite now made more sense: it might be that an exceptional diversity of ancestry was behind the prevalence of mutations in Colombia. Kosik was starting to think that it couldn't be coincidence that so many presenilin mutations should coexist in this corner of the world. The simplest explanation, he said, was that Colombia, during the Spanish Conquest, saw a cornucopia of rare mutations from African, European, and Indigenous sources. It also suffered, during that same period, a number of genetic bottlenecks. In a bottleneck, a population shrinks due to some disaster—disease, war, genocide, or famine—leaving the surviving population with reduced genetic variation. If and when that population bounces back, rare variants that made it through the bottleneck become overrepresented.

Lopera wasn't up to speed about any of this. I'd just heard him repeat to Randall Bateman that Antioquia was a genetic isolate, and he told me recently that he expected that all the newly discovered mutations would also prove to be European in origin. His attention was consumed with running the API Colombia trial and the possibility of a DIAN trial, not questions about genes and ancestry.

But Kosik was enthralled. Immediately after returning to California, he and his doctoral student Juliana Acosta-Uribe set out to discover the origins of the Girardota mutation. They compared samples from the family with reference data from the 1000 Genomes Project, a catalog that scientists mined to discover DNA haplotypes from populations all over the world. The Girardota family, it turned out, had both European and African haplotypes in their individual genomes, as well as a smaller share of Amerindian ones. After a few months, Acosta-Uribe narrowed down the haplotype where the I416T mutation occurred and was able to take a closer look.

The 1000 Genomes Project registered twenty-five people with the same haplotype, most of whom could trace their origins to the modern-day nation of the Gambia, in West Africa. No one knew exactly where or when the mutation had emerged. None of the reference samples from the Gambia had it, meaning that it, like E280A, may well have arisen in Colombia. What was clear was that it had emerged in a person of African descent. This paisa mutation was African.

ON FRIDAYS AT the SIU, Laura Ramírez, a neurology resident, conducted medical evaluations of the Girardota family members, while Sonia Moreno, the neuropsychologist, performed cognitive tests. Along with Kosik's team and other collaborators, they were planning a paper that would bring together all the genetic, imaging, and cognitive findings from the family. If any of its sick members should die in the meantime, their autopsy results—the Girardota family's first—could be included as well.

That day, Ramírez was evaluating Alex, an athletic twenty-nine-year-old who radiated good health, though his mother, I learned, was now bedridden with Alzheimer's. Alex had skipped his last year of high school in a bid to play professional soccer, but once that opportunity dried up, he worked and studied photography. He did not drink or smoke, he told Ramírez, and he avoided eating meat. Lately, he reported, he was coping with insomnia. And he was starting to forget things.

Ramírez pressed him to explain about the forgetting. He would rate his memory an eight out of ten, he said, where a year ago it was a nine.

Alex's concern surprised me; he was more than a decade away from the first symptoms of the disease, if he was fated to develop it at all. Ramírez reassured him that his memory wasn't failing and advised him to stop looking at his phone late at night. "People in these families panic over little lapses in memory," she told me after Alex walked out. She hated terms like *picado*—with which the families tagged someone they thought was becoming ill—and knew how they put people on edge.

Ramírez's next patient was Alex's uncle Álvaro, a man with evident dementia. Álvaro, in contrast to his nephew, had zero years of education and had never learned to read. A lifelong construction worker, he was far too impaired to work. He could no longer leave the house without getting lost or be trusted to handle money, even coins. Before meeting with a patient, the investigators customarily interviewed a spouse or caretaker in private. Álvaro's wife, a depleted-looking woman in her fifties, entered the consulting room first, alone.

Álvaro could still bathe himself and ate normally, she reported, but he sometimes forgot to do either, and when he was reminded he got angry. They had been arguing regularly—fighting over something she had or hadn't said—until she learned it was easier to stop quarreling and retreat into a self-protective silence. He refused to take the antipsychotic drugs he'd been prescribed to keep him calm, and he stayed up all night drinking coffee. Álvaro had lost his taste for alcohol, even though he'd previously drank so much that he was hospitalized with cirrhosis.

Ramírez summoned Álvaro in. He was a lean, muscular man in a senselessly blissful mood, twitching and moving his head as he answered her questions.

"Why are you up all night?" she asked him.

"I'm thinking and thinking and thinking," he responded. But he wanted to sleep through the night and "rest my brain," he said, still smiling. Álvaro walked stiffly to the exam table, where Ramírez checked his reflexes and looked into his eyes. People's eyes changed over the course of their disease, with a gaze that became progressively removed. Many clinicians felt they could detect something in the eyes before any other symptoms presented. Even those who rejected a cruel term like *picado* harbored private hunches about the disease, born of experience and observation.

When Moreno invited Álvaro into another room for tests, his wife asked Ramírez whether Álvaro would end up like his siblings, a few of whom by now had died. She wanted to be able to collect what little pension her husband was entitled to. And she wanted to know if Álvaro carried the mutation.

Ramírez answered that she was not privy to knowledge of patients' genotypes, and that families were not told either. But it appeared likely, she advised the woman, that Álvaro had the family disease and would only continue to get worse.

Álvaro's wife nodded numbly. She was unkempt in a region where women are rarely unkempt, with dirty fingernails and unbrushed hair. The burden of her husband's disease seemed to have undone her, even as Álvaro—still walking, talking, and eating unaided—was in the early stages.

The couple left. As affecting a picture as Álvaro and his wife presented, it was their young nephew's angst that haunted me more. Alex's plight reminded me of Daniela's, and I found myself flushed with indignation: that you could make every effort to build yourself a better life than that of previous generations, and still end up just as sick.

A MONTH AFTER Randall Bateman returned to St. Louis, discussions about a trial recruiting the Girardota family were still being held up over the genetic counseling issue. Lopera had insisted that counseling was not standard in Colombia, but Bateman's group had conducted its own investigation and determined that counseling and testing were in fact available, and for many diseases they were already standard.

Having worked with early-onset Alzheimer's families around the world, Bateman knew that people had a broad landscape of attitudes when it came to learning whether they were mutation carriers. The DIAN investigators' experience was that most wanted at least the option to know. A smaller number then went ahead with genetic counseling. People who did not want to know were discouraged from joining clinical trials, lest they develop a side effect, such as nausea or a skin reaction, that revealed them to be receiving the active drug—which only mutation carriers got. For Bateman, offering genetic information before a drug trial was both an ethical and a safety issue on which he was not inclined to budge. "There's no way for us to say, 'For everybody else we think this is necessary to offer, but for Colombia we don't.' It's hard to justify that," he said.

This made sense, enough so that it seemed odd that the API trial had proceeded as it did, risking the emergence of side effects that would expose people as carriers. Bateman said he'd been on the phone a lot recently with Eric Reiman, Ken Kosik, and others, trying to figure out some way to break the stalemate with Lopera. "My perception so far has been that the biggest concern is the concern of the unknown," he said. "Sometimes you just don't want anything to change because you just don't know what's going to happen."

FOURTEEN

Two months after her mother's death, Daniela sent me tearful WhatsApp messages from Bogotá, distraught at what she deemed a lack of information from Lopera's team. Why had she and her sister never received a printout of their mother's autopsy report, as Andrés Villegas had promised them? What was Neurociencias really doing with their mother's tissues? Her voice messages were long, and she spoke so fast I had to listen to them repeatedly. She was upset about a lot of things, all at once. She had recently tried and failed to secure a job with Bogotá's transit system, and she was worried about the country's upcoming presidential election. Daniela and her partner, Ximena, feared that their candidate, the leftist former senator and Bogotá mayor Gustavo Petro, would be robbed of a victory by the machinations of the extreme right. I struggled to keep track of her complaints and accusations. All that came through was her pain.

Now Daniela was essentially a housewife, a role for which she had all the skills but no appetite. She bathed and fed Ximena's daughter, shopped, washed, mopped, and made stews in her pressure cooker. The couple lived in a suburb bordering a protected lagoon and forest, where Daniela gathered the medicinal herbs that Doralba had taught her to use: bitter rue for a bath, a spray of elderflowers for a tincture to soothe a cough. Ximena, a bookish, engaging woman in her early thirties, worked as an editor. She liked to go out with her colleagues and often didn't return until late at night, leaving Daniela home all day to mourn Doralba and ruminate.

I'd mentioned to Daniela the strange reference to an owl from her mother's clinical history, and the story haunted her. Doralba was just thirty-three years old then. Was her disease causing her to hallucinate already? If so why, Daniela wanted to know, couldn't something have been done for her sooner? If Lopera suspected then that Doralba was becoming sick, why didn't he start her right away on the transdermal patches of rivastigmine that the Alzheimer's patients got? The medication didn't do much, Daniela knew, but at least it would have been something.

Daniela had long felt that the researchers put their own interests first. She remembered a program Neurociencias had run more than ten years before, when they asked siblings from a few selected families, including the López Pinedas, to play games and take tests. The ones who participated—including her mother and three of her aunts—had since gotten sick. This led her to conclude that the staff must have known all along who among them carried the mutation. It wasn't true; only a tiny number of people in the research group had access to that information, and only in specific circumstances could they see it. But Daniela believed otherwise. "They do a battery of tests. They never give results. They just watch how all this unfolds in every patient," she complained.

When Doralba was finally diagnosed with dementia, Daniela approached Lopera, asking to learn if she, too, was a carrier. He told her she needed to wait until the API Colombia trial had ended, after which everyone could seek their genetic status. By then, she calculated, she'd be thirty years old.

LAS FAMILIAS

OVER XIMENA'S OBJECTIONS, Daniela boarded a bus and headed to her sister's place in Medellín, not saying when she would be back. It was just too much for her to be away when the wounds were so raw, and when there were certain types of tragedies, and patterns of experience, that only a family like hers could understand. The López Pinedas were like hundreds of E280A families under the lens of Neurociencias: city people with country roots who struggled to attain personal and financial stability, their efforts complicated further by generational waves of illness and death. Most of Daniela and La Flaca's cousins had already lost a parent, or were soon to.

Daniela invited me to lunch at La Flaca's, along with María Elena, their youngest aunt.

La Flaca and her husband, Fernay, had carved out a fragile middle-class life in Comuna 13, in a better sector than where she and Daniela were raised. Fernay drove a cab, and Flaca bought clothes wholesale from sweatshops and resold them to family and friends. The couple's apartment was like many homes I'd seen in Medellín: sparsely furnished and scrupulously clean. Cleanliness here was like a type of wealth, I thought, or a substitute for it: the simpler the house, the shinier the surfaces, the stronger the smell of detergent.

Half sisters who were closer than most full siblings, Flaca and Daniela sat on each other's laps and finished each other's sentences. Doralba had been the center of their world, and now they relied on each other.

They talked about Alzheimer's disease often. It touched their daily lives so much that they had to. It was like a family business, a niche enterprise whose terminology they needed no glossary for. In recent days Daniela had reiterated her desire to be tested for E280A, but Flaca opposed this—so much so that she'd started lobbying other family members, including María Elena, to try to talk Daniela out of it. Flaca's argument was that until there was a cure or at least a meaningful treatment, nothing good could come of knowing. She didn't think Daniela, in the state she'd been in since Doralba's death, was psychologically prepared for a bad result. The sisters bickered over lunch until Fernay carped at them to stop. "Neither one of you has it," he said. I didn't know whether he believed that or was just trying to be kind.

María Elena arrived after we'd all finished eating. With her new Samsung phone and designer sneakers, she gave off a more prosperous air than her nieces, whom she presented with gifts of red lipstick. She settled herself into Flaca's black leather couch and wasted no time upbraiding Daniela about wanting a genetic test.

"You're a masochist," María Elena said.

"I'm not," Daniela protested. "I just want scientific proof. For myself."

"Look," María Elena said, now addressing me. "There are things that are hurtful to know, and she wants to know them. It's like, if your husband has a girlfriend, going and looking for her."

"That is masochism," La Flaca agreed.

"Masochism is me thinking my husband is cheating on me, and I'm jealous, so I follow him and confront him," María Elena continued. "Who's the one that's going to suffer? Me. The other woman won't suffer and neither will he, and he probably won't leave her. So why should I want to know something that's just going to bring me problems? It doesn't help."

Daniela listened with polite resignation as her aunt spoke. Nothing María Elena said seemed strange to me, but her eyes were sunken and still, not matching the expressiveness of her speech. "I feel good, and I just turned forty-four," she went on. "My brothers and sisters got sick much sooner than that. I don't want to know anything. I just want to keep living. Anyway, I always say that I'm destined to be healthy, because it can't be possible that there should be so many sick people in my family."

Though she had not joined the crenezumab trial and wanted no part of any genetic test, María Elena still lent herself generously to Neurociencias. She arranged for the brains of her siblings to be donated when they died. She showed up for testing whenever she was asked to. In recent months she'd traveled to Boston, with a group that included some of her cousins from Angostura, for brain imaging. She recalled performing memory tests while lying in an MRI tube, walking on a treadmill while breathing oxygen from a mask, and solving arithmetic problems while the investigators made her distinguish between similar smells.

I told María Elena I couldn't believe she would do all that and not want to know if she carried the mutation.

"I don't want to think about it. I just want to help, until there's a vaccine, or a cure, and until then I'm there for whatever they need," she said.

María Elena showed us recent photos of her sister Elcy, who was now in a wheelchair at a city-funded nursing home. She'd just been to see her sick brother Fredy, who the year before could only talk about the past, about his time in the army. He had since stopped talking.

Daniela, Flaca, and Fernay took this all in knowingly.

"Their old memories come back, but just for a while," Fernay said.

"It comes and goes," said Daniela. "It's when they're falling into that trance of old memories that they start to lose new memories and then begin to lose their speech."

The López Pinedas saw Alzheimer's differently from how most people did. To them it was not simply a disease that erased memories, but one that changed the mind and body in patterns that varied from person to person. They never tired of reviewing the cases in their family, the specific symptom trajectories of each.

María Elena went on to report on yet another sibling: her sister Mabilia had reached the "fat stage" of the disease, she said. This was something they all recognized: Alzheimer's patients can go through a period of excessive eating, when the brain no longer registers fullness. It had happened to Doralba, too.

If there was one thing she'd learned in all her time with Neurociencias, María Elena said, it was the centrality of the brain. "It's your heart and soul and your mind. It's your motor. When it starts to atrophy, it affects your movements, your speech, your weight," she said. "Everything."

WALKING WITH DANIELA a few days later, I recalled what her aunt had said about getting tested for E280A—that it was like stalking an unfaithful husband. It wasn't just a metaphor, Daniela told me. María Elena's husband had maintained a second family in secret for twenty years before she found out.

Right, I said. Well, besides that, María Elena had seemed fine to me.

The illness can seem to go dormant for a while, Daniela said. Sometimes, after an initial burst of symptoms, nothing might happen again for a few years. She hoped her aunt didn't have the disease, but she had a feeling about it that wouldn't go away.

We were on our way to see her very sick aunt Mabilia, who lived in Pedregal, the neighborhood where the López Pinedas had settled after arriving from Angostura. Mabilia's home was a small brick box whose front door opened into a living room. A woman let us in—it was Mabilia's sister-in-law Gloria, who had nursed one of Daniela's uncles until his death. Now she was with Mabilia every day, while Mabilia's son took over at night. Mabilia's husband, Checho, drank all the time, Daniela told me, and did absolutely nothing to help.

As we entered, I saw Mabilia, a woman with a gray stringy ponytail sitting in a plastic chair with her back to us, turning her head slowly left to right. The room was dark, and it took me a moment to realize that she was facing a mirror, staring happily at herself.

Daniela seemed to crumple at the sight of her aunt, whom she hugged and kissed as her eyes welled up. Mabilia responded with delight, making kissing noises. Her fingernails were painted, something her sister-in-law had taken it upon herself to do. Daniela told her aunt how pretty she was. Mabilia returned the compliment. "Pretty," she said back.

Daniela spoke clearly and repeatedly to Mabilia, as one would with a child. "I'm Daniela," she told her aunt, cheerfully. "Dan-ie-la." Would she like a juice? An ice pop? Daniela had ample experience getting through to a person with dementia, using simple language, keeping her tone playful, repeating herself as often as she needed to without a hint of frustration.

Mabilia assented to an ice pop and Daniela ducked out to the store.

I sat with Mabilia in the darkened living room as she admired her reflection over and again, rhythmically turning her head and smiling. A Bible sat open on a stand in the corner, just like in Flaca's house, and country music played. Mabilia did not acknowledge me, nor did I know, the way Daniela did, how to communicate with her.

Daniela returned with the ice pop, now composed and cheerful. She reciprocated her aunt's kissy faces, joked with her. Was Mabilia's husband Checho handsome or ugly? she asked. Handsome? Ugly? Checho?

It took Mabilia a few tries to respond. The word she wanted seemed lodged in her mouth.

"Ugly," she finally burst out, and they both laughed.

VALLEY OF FORGETTING

THE NEUROCIENCIAS CLINICIANS knew they were not universally loved by the E280A families. Even in the same household there might be admirers and detractors. This was the case with the López Pinedas of Daniela's generation. While Daniela was somewhat skeptical of the researchers, Flaca welcomed all contact with them. She had missed the age cutoff for the crenezumab trial, but hoped to become eligible for another trial down the road. She had recently completed a course as a nail tech courtesy of Neurociencias. It was the latest offering to the E280A families under the auspices of its social support program: free training in barbering for men, and in nail care for women.

On the hostile end of the spectrum was their first cousin Lina. An olive-skinned, petite woman in her early thirties, Lina, too, had lost a mother to E280A and was at risk for it herself. Lina was like another sister to Flaca and Daniela, but more focused and self-assured. She wanted nothing more to do with the researchers, with the exception of Francisco Piedrahita, "because he's one of us, so he understands what we live through," she told me. The rest, as far as she was concerned, could go to hell.

Why? I asked her when we met, while she took her lunch break from the pharmacy where she worked. In the space of an hour, with obscenities flying and zero sentimentality, she explained how she had become alienated from them during the years she strained valiantly to care for her dying mother. Complicating matters further, she said, it was a mother who had never shown her love.

Lina had never been able to confirm that she had been conceived from rape, as her mother, Amparo, had always said. But Amparo's utter indifference to Lina made Lina think it was probably true.

When Lina was a young child, Amparo worked as a live-in nanny during the week, and visited some weekends. Lina lived with her uncle

Jaime and her grandmother, Roselia, in the family home in Pedregal. Her mother's other siblings—Fredy, Doralba, Elcy, Mabilia, María Elena—lived on and off in the same home in those years, when the López Pinedas were fresh from the countryside and struggling. This worked out well until Roselia died of Alzheimer's disease, in 1992. Lina was eight years old then, and her world changed forever. No longer protected by her grandmother—who so favored Lina that María Elena, her youngest aunt, cried foul—Lina now had to learn to wash her own clothes. After Roselia's death, Jaime married, and his new wife set about emptying the home of all the brothers and sisters and their children whom she considered hangers-on.

Lina briefly went to live with her mother, who was in a new relationship and pregnant. But the man left Amparo after Lina's sister was born. Their refrigerator was repossessed, their electricity was cut off; soon there was no food. Amparo went to work in another home, taking the baby with her, and Lina was sent to live with some cousins on the Pineda side. It was the type of arrangement many poor mothers made, leaving their children in the care of relatives, sometimes paying them and sometimes not. Lina spent her adolescence in that household, where her second cousin, a grown man, touched her and spied on her repeatedly while she was changing. When she finally confronted him, making a scene, no one thought what he'd done was weird. Instead the women maligned Lina further. They hid her pain pills when she had her period, and made sure no pads were in sight.

She survived in part thanks to her aunt María Elena, who snuck her the necessities that the cousins denied her. She emerged an organized, studious teenager, but one who radiated anger. When Lina graduated valedictorian of her high school, no one came to witness or celebrate her achievement. She resented having a mother who had left her to the mercies of a pervert and his enabling family. "Why didn't you abort me if you were just gonna give me this kind of life?" she demanded of Amparo. Daniela and La Flaca, with whom she was close growing up, were also poor. But at least they were mothered.

After high school Lina sold lottery tickets, socks, and magazines until she encountered an ad for a program at the public vocational college. The school was one of the few ladders available to young people of meager means. Getting in involved making countless phone calls from midnight on the day of registration, having the random luck to receive a code, showing up with that code at a specified day and hour, and finally a series of tests and interviews. "I dusted off my notebooks and my stubby pencils," Lina said, and applied for the pharmacy program. Her cousins laughed at her as she left the house dragging her old school bag, but she was admitted. They let Lina know they would not pay her bus fare; all she could expect was breakfast, lunch, and a roof over her head. So Lina began walking or even jogging to class and back, more than an hour each way.

Her uncle Jaime, after hearing about Lina's plight, offered to pay her bus fare, her supplies, and her uniform. María Elena chipped in for shoes.

And your mother? I asked. She didn't help?

"She told people she was proud of me," Lina said. "But no, nothing. If my sister needed something, she'd hold a raffle. She'd sell empanadas if she had to. But for me, no."

Lina knew that her mother—along with Doralba and her other siblings—had taken part in Neurociencias studies since the 1990s. Lina never understood why Amparo participated so faithfully. The cash she received wasn't much. The threat of Alzheimer's disease was still distant and opaque. Lina's grandmother's symptoms had become evident only a few years before she died, and there was not a sense in the family, as there would be later, of a dark cloud threatening to envelop them all.

When Lina was in her twenties, working as an assistant pharmacist, her mother's employer called to report that Amparo had been acting strange—she had put powdered incense in their food—and they were letting her go.

Amparo begged Lina to make room for her and Lina's younger sister. Though she was barely able to feed them with her pay, Lina was happy to have her family under one roof again, to be the provider she'd always dreamed of being. It remained unclear to her whether anything was really wrong with her mother, who still spoke and cooked well, despite the incident with the incense. "I would go to work and my sister to school and she stayed home," Lina said. Within two years, however, Amparo "started having her little moments," Lina recalled. "She would make us laugh with the silly things she said."

Lina still remembers the date her mother disappeared: May 19, 2011. She assembled a search team of her cousins and friends and called the police and fire department. After a few fruitless days she called the local TV station and the tabloid *Q'hubo*. The staff of Neurociencias were aware by then that Amparo was missing, but they did not take part in the search effort. "And that's why I'm so horrible with them," Lina said. "For not having helped while my mother was always their guinea pig. They needed an MRI, there was Amparo; they had a meeting, there was Amparo; a blood draw, there was Amparo; a cognitive test, there was Amparo. For everything, she was there. And in her moment of need they weren't there for her."

On the seventh day—well after most of the López Pinedas had quietly given her up for dead—Amparo resurfaced. When the figure of a woman was spotted in a creek far from where Amparo and Lina now lived, someone called the police, who found Amparo unconscious and wounded, but alive. When *Q'hubo* got the news, they called Lina, demanding an interview in exchange for information on where she could find her mother, and Lina assented.

At the hospital, Lina learned that Amparo had been beaten, burned, and possibly raped. "It was very confusing," Lina recalled. "She was in shock and delirious. She said she had been taken into a house by several men. Later she said she was never in a house. We didn't know if she had been raped again or if she was remembering her past."

After that, Amparo could no longer be left alone. Lina beseeched her aunts and her cousins to help her find a safe place for her mother, commencing a series of moves to different homes and neighborhoods that would go on for years. Later Lina rented a studio apartment where she locked her mother inside during the day, making sure to turn off the gas before she left for work. Her younger sister refused to help, once delivering a disoriented Amparo to the door of Lina's pharmacy to be rid of her. When the sister turned eighteen, Lina took her to court to force her to help support Amparo, a move that shocked the whole family. "If you have even one potato to eat, you're going to give half that potato to your mother," Lina warned her.

Eventually Lina had to seek a nursing home for Amparo. Her landlady was pressuring her to get her mother out, fearful she would cause an accident. Lina searched all over Medellín and its outskirts for a home. None would accept patients younger than fifty. She approached Neurociencias to see if they could help. "And you know what Lopera told me? That I should start a fund with my family. I told them to eat shit—that they would never have anything to do with my mother again."

One reason a family fund was out of the question was that Amparo's siblings Fredy, Nicolás, and Elcy had all fallen ill, and Doralba now appeared to be in the first stages as well. The López Pinedas were a family in crisis. A disease barely on their radar just a few years before was now claiming five of its members. Lina had canvassed every social services agency in Medellín and was at the point of mental and physical collapse when she learned of a nursing home run by nuns. She did her best to explain her plight to them, and though they listened, they told her the same thing all the others had: there were no beds. "I was defeated," she said. "I wanted to die. 'This is fucked—I'm fucked,' I thought."

Before Lina could leave, however, she felt compelled to beg the nuns one more time, her voice cracking and her lips trembling as she told them how sick her mother was—and how close she herself felt to dying.

The nuns conferred. The issue, it turned out, was that they literally had no bed. They did, however, have space in a corner. If Lina could buy a bed for Amparo, they would care for her there. She would have to pay a little bit every month, and provide all her mother's nutritional drinks, as well as her diapers and creams and soaps. Lina accepted; it was a pact that each side honored for the remainder of Amparo's life.

Once Amparo was safely under the nuns' care, Lina barred Neurociencias, and her own family, from visiting. She'd shouldered the burden of her mother's illness without their help, and she was uninclined to accommodate them now. "My family called me selfish," she recalled. "I said, you *hijueputas*—what do you want to visit for? Just to gossip? To watch this drama unfold? Forget it." Only when Amparo was in her death throes, after a priest had administered last rites, did Lina allow two of her mother's sisters to say their goodbyes. When Amparo died, her brain was buried with her.

FIVE YEARS HAD PASSED since then and Lina did not come off as angry, at least not to me. She had felt the cloud of resentment and rage—"which in the end was really just despair," she said—start to lift after her mother's death. She met her partner, a systems engineer and professor with whom she went on to have a son. She repaired her relationship with her sister, who was married with a child of her own. She even forgave, and kept in touch with, the second cousins with whom she had lived. The cousin who had touched and spied on her was very ill with Alzheimer's, as was his sister who had defended him.

For a while after her mother's death, Lina thought she wanted to learn if she carried the E280A mutation. When she approached Lopera, though, he explained the test to her as "something you could only do in Phoenix, Arizona, and places like that," she said. Why not enjoy her life instead? Lopera asked. A bad result could drive her to suicide, he warned. Lina insisted that she didn't care.

"But when my son was born, that changed," she told me. "Now I don't want to know."

Neurociencias had called Lina about enrolling in the crenezumab trial. It relieved her to know that they didn't seem to hold her prior pugnaciousness against her, but she demurred—at the time she was nursing her baby—and they hadn't called since. Lina feared for her aunt María Elena, whom she felt was on the brink of becoming sick, and for herself. She was just ten years younger than María Elena. Birthdays felt ominous to her, and she dreaded each one. But most of the time she tried not to think about it.

"I don't visit my aunts and uncles who are sick," she told me. Not even the ones she loved. She did call her cousins to offer support, and to let them know that if Fredy or Mabilia needed something—food, diapers, anything—she would always be there for them, because she knew how painful this disease was, and how isolating and impoverishing. But she asked them all to understand that she could not and would not visit, because each time she looked into their withering, innocent faces, she could see only her mother.

FIFTEEN

In the spring of 2018, the API Colombia team published its first major scientific paper, in the journal *Alzheimer's & Dementia*. The paper described the E280A mutation cohort, the study drug crenezumab, the trial's design that let both carriers and noncarriers participate, and the outcomes being measured. While results were years away, the paper nonetheless signaled a triumph: the successful launch of an important Alzheimer's trial in a place where few thought it was possible, largely thanks to the efforts of Francisco Lopera and his colleagues.

So it came as a shock to the team when Lopera was listed neither as the paper's first nor last author, the two names of greatest consequence on any scientific paper. The first author named was Pierre Tariot, the last Eric Reiman—the directors of the Banner Alzheimer's Institute.

Ken Kosik felt sure that they had put the authorship question to Lopera, and that Lopera had somehow assented. Nevertheless, he felt the move was wrong, and reeked of the sort of scientific imperialism he and Lopera had strained for decades to avoid. "Lopera spent thirty years doing this and now he's listed as second author? This is his life's work on the line," Kosik told me from Santa Barbara, where he took it upon himself to craft an angry email to the Banner scientists. Though moderate in tone by any normal standard, for a neurologist it was a fiery shot across the bow.

"We have discussed many times the pitfalls of conducting a clinical trial in a developing country, particularly the appearance of exploiting a poor and vulnerable population," Kosik wrote. "To make it clear that the API project is not exploiting the local population for the benefit of yourselves or a pharmacologic company sponsor, it is both ethical and to everyone's benefit that a member of the Colombian team be given the intellectual recognition they deserve in this paper."

None of this could have occurred, Kosik continued, "without three decades of work by Dr. Lopera traveling to remote villages throughout Antioquia at a time when travel in the region was fraught with danger. The work in the paper represents countless hours of collecting family trees, doing neurological exams, and finally demonstrating that the pathology was [Alzheimer's] and identifying the responsible gene. His work did not stop there. His ongoing effort in enrolling patients should be appropriately recognized."

Sent. Waiting for death threats, he reported the next morning.

A week later Kosik arrived in Colombia, where he and Lopera traveled to the Caribbean beach towns where Lopera had worked in the 1970s. They swam, snorkeled, and ran into people from Lopera's time there, including a woman he had successfully treated through a high-risk pregnancy. The trip was meant as a vacation, and neither Kosik nor Lopera felt much like talking about the *Alzheimer's & Dementia* paper. But Lopera, who had said nothing to the Banner scientists, was glad to know Kosik had.

Far from defiant, Reiman and Tariot responded politely to Kosik's letter. Weeks later, in Arizona, Tariot told me that API's ethics and cultural sensitivities committee, which had helped shape the trial's design, would "join us in this conversation about—with an international collaboration like this—what's the right way to give credit."

I asked what would happen when the final trial results were published. "I think if I were to predict, I would say that major papers have to identify Francisco as the leader of the trial effort there," Tariot said.

Something that struck me in Kosik's letter was his argument that crediting Colombian investigators was a way to inoculate the researchers against any allegations of exploiting poor people in a developing country. Did that mean that Lopera and the families were to be considered of a piece, I asked Kosik, and that an ethical debt paid to one sufficed to cover the other? "It's a reasonable question," he conceded, but he didn't have an answer. Kosik was used to thinking about what the Colombian investigators were owed, more than the families.

The Neurociencias team often referred to themselves and the Alzheimer's clans as "one family." With the exception of individuals like Lina, whose experiences with the research group were bad, the family members tended to hold Lopera in high esteem, and Lopera viewed himself as their protector, negotiating their dealings with the media and scrutinizing all proposed experiments or studies involving them, quickly nixing any that appeared pointless or harmful. The benefits of the crenezumab trial were mutual: Lopera got the distinction and support of heading a major clinical trial, while the families got access to a drug that might stave off a dreaded illness, along with payments that, while trivial by first-world standards, were not trivial to them. But I wondered whether, over time, the balance of interests between the researchers and the families could change, especially if more players—foreign research institutes and drug companies—got involved with Neurociencias.

IN JULY, LOPERA, his team, and a few members of the Colombian families traveled to the annual Alzheimer's Association conference in

Chicago, where news from other drug trials would be presented. Pharmaceutical companies had largely staked their Alzheimer's programs on the amyloid hypothesis, whose adherents saw amyloid-beta as the fire in the brain that needed to be put out. Detractors, by contrast, saw amyloid as the smoke.

Scientists were exploring dozens of other pathways that could lead to brain damage in Alzheimer's disease. Among the potential culprits were misfolded tau protein, whose spread closely correlated with symptoms and was implicated in other types of dementia; abnormalities in microglia, a type of immune cell in the central nervous system; and disruptions to the lipid metabolism of neurons. But the big money was on amyloid-beta. A host of experimental drugs had been shown to attack the protein in its various forms—before, during, and after its aggregation into plaques—and clear it from the brain. The question was whether doing so would slow people's march to dementia.

Critics of the amyloid hypothesis noted that it had been largely generated from observations of familial Alzheimer's patients, in whom amyloid-beta featured prominently. The first brains of the E280A patients, studied at Harvard, had been full of amyloid-beta 42. Such findings may have dulled researchers to the possibility that there were other pathways that could be sparking the neurodegenerative process. But if anti-amyloid therapies were to work in anyone, early-onset families were the ideal populations in which to find out. Besides sharing the same mutation, diet, and way of life, many of API Colombia's participants had no plaques in their brains at all when they enrolled. They offered one of the last, best tests of the hypothesis to date—as long as the drug reached their brains.

Though API's infusion center finally opened that summer, Colombian regulators still had not signed off on the high-dose infusions that the scientists felt was their only hope for success. Five years into the trial, participants were still getting too little crenezumab.

LAS FAMILIAS

WHILE LOPERA AND his team were away in Chicago, I pored over the López Pineda genealogy at the SIU. The family belonged to the C2 clan, which was concentrated in Angostura and its hamlets. The largest of all the E280A families, C2 comprised some two thousand people—roughly a third of the cohort. The thick, dusty file dated from 1999, the last time these genealogies were regularly printed out. The document was covered in corrections made in ballpoint pen: *Did not get sick. Got hit by a car.* Some of the notes marked the discovery of entire new branches, superimposed haphazardly onto the original C2 tree by hand. As the generations progressed, black squares and circles marking illness became less frequent, as the younger members had yet to display symptoms. Daniela and Flaca and Lina were only seven, thirteen, and fifteen years old when the document was created, three little white circles attached to fathers marked as *NN*: unknown. From their grandmother, Roselia Pineda, their lineage traced easily to the couple considered the originators of E280A: José Miguel Pineda and Petronila Sampedro, the parents of the sisters who had colonized the hill towns more than two centuries before.

XIMENA WAS PRESSURING DANIELA to return to Bogotá, but before Daniela did, she wanted to see the house where her mother was raised and meet the country branches of their family, the Lópezes and the Pinedas. Her mourning of Doralba had entered a new phase. The anger had dissolved, replaced by an inquisitive urge. She wanted to learn things about her mother, to talk with people who had known her. Neither Daniela nor her sister had been to Angostura since they were girls. María Elena advised them that the López clan would surely welcome them, but the Pinedas would not. That branch of the family treated their disease as a secret. They'd hidden their father when he became ill with Alzheimer's, and were now said to be hiding one of their brothers. Neurociencias could never coax them into studies.

Daniela, Flaca, Flaca's son, and I left by bus the next Friday. We passed through Yarumal, then the tavern at Canoas, until a giant roadside monument to Padre Marianito announced that we had arrived in Angostura, where Padre Marianito's sepia-colored portrait stared at us from every wall.

We checked into the Hotel Los Recuerdos, the Memories Hotel, a simple and clean place on the plaza. Daniela wasted no time interrogating a group of old men gathered on the steps of the church, watching buses and chivas come and go. Did anybody know, she asked them, where she could find the home of Abel the tailor?

Abel the tailor had been dead for forty-some years and yet no one found the question to be strange. An old man with a Marianito medal chained to his belt loop told us where to find Abel's former house and walked us halfway there.

The house turned out to be enormous. A family lived there now, and a girl of about ten, in a peasant skirt, gave us a tour. Inside the home's thick earthen walls it was cool and pleasant. Daniela touched its ancient doorframes and granite washbasin with fascination and reverence. She imagined Doralba, the tailor's daughter, walking down this steep street to the base of the plaza to bring Abel his lunch, as Doralba had recalled to her years before. The young girl explained to us that her grandfather had bought the home. "And there on the street is where they killed him," she said, pointing. We asked no more questions.

That night, as Flaca and her son retired to their room and rainwater made torrents in Angostura's streets, Daniela plowed on, looking for the López cousins. María Elena had given her a vague idea of where they lived, but it was the sight of a young girl with a domed forehead, sitting inside a doorway, that confirmed it. "You see that forehead? She's one of us," said Daniela, who turned out to be right.

The Lópezes graciously let us in despite our appearing out of nowhere in the evening downpour. Their patriarch was Javier, an elegantly dressed man in his late eighties, who lived with his sons and their families. Javier's house was humble but spacious, with a back garden full of dogs and chickens. Javier, like his brother Abel, had been a tailor. An old photo on the wall showed Javier alongside both of his brothers, then in their twenties and thirties, with slicked-back hair and mustaches, looking like three rakes about to go conquer a tango bar. This was the first image of Abel that Daniela had ever seen.

Javier's son William worked in a hydroelectric plant that had been bombed twice by guerillas, while another son, Carlos, managed a feed store. A third brother of theirs, now dead, had been close with Doralba, they told us, and they pointed out his photograph on the cracked living room wall among their forbears and their saints. It was an oft-displayed image in Colombian homes: the murdered father, brother, son.

The López family was aware of the genetic Alzheimer's in Angostura and of all the investigations that had transpired here. Javier's son Carlos had been working as a guard in Angostura's hospital when the Neurociencias team performed its first brain autopsy, back in 1995. But they talked about it as if from a distance. They didn't consider it to be their disease so much as the Pineda clan's. It had come into their family twice through marriage, first when Abel López married Rosalia Pineda and gave birth to Doralba and the other López Pineda children. More recently, one of Javier's daughters had married a Pineda son named Diego. That marriage had failed, the daughter left for the United States with her child, and now Diego was sick with Alzheimer's. This left Javier's grandchild with the chance of being a carrier.

Daniela ducked her relatives' questions about her personal life as they served us chicken and rice. Instead she recounted more of her family's odyssey with Alzheimer's. They didn't entirely understand the disease—was it just that you lost your memory? Daniela gave them her capsule version: You begin to lose memory and start acting strangely as your neurons start to die, she explained, and the neurons go on dying until you're bedridden, no longer able to do anything for yourself.

When the rain had diminished to a drizzle we said goodbye to the Lópezes, who said they expected us back the next day. We made our way to the Hotel Los Recuerdos as men with machetes on their belts stumbled drunkenly around the wet stones of the plaza.

THE CHURCH BELLS STARTED at 6:30 the next morning. I woke up early to walk with Daniela, who left the hotel in a Che Guevara T-shirt, but went back inside to change after it attracted startled looks in the plaza. The last thing you wanted to look like in Angostura was a Marxist.

Compared to the lonely hill towns of southern Spain, which it resembled on the surface, Angostura struck me as a living, pulsating remnant of medieval Europe. As I waited for Daniela, the plaza began to fill with farmers bringing in wobbly calves and bags of fat carrots to sell. The *arrieros*, the handsome cowboys of these hills, led in their endless trains of mules bearing heavy packets of cane sugar, as the priests ascended the steps of the church in their flowing white robes. Soon Uncle Javier would emerge from his home in a collared shirt and felt hat and head to the steps of the church, as though he were leaving for a day's work.

The town was small. Where the Lópezes' street dead-ended in pasture, Daniela and I clambered up muddy mule trails, some of them narrow canyons taller than our heads. We slid under barbed wire fences until we reached a hilltop, from which we could make out all of Angostura: the forests, the rushing creeks, the town that appeared to be laid out like a cross. Away from the critical eye of La Flaca, Daniela smoked a half a joint she had saved as we rested on the grass.

The Lópezes had talked a lot about Padre Marianito's miracles and also about witchcraft, two things that Daniela believed were related. How was it not witchcraft to steal a piece of a dead priest's finger, as people here had done? It was common for country people to practice magic, she said. Before his conversion to Pentecostalism, her stepfather, Orlando, used to do things like put salt on his face, fill socks with herbs, and tie up dolls.

The balcony of the Hotel Los Recuerdos looked out at the facade of Angostura's church, on which hung a giant portrait of the young Marianito—looking downright intimidating, with one eyebrow cocked warily and black eyes focused in an intense, reckoning gaze. A trio of shops by the church sold thousands of Marianito products, from life-size garden statues to mugs and keychains.

As Daniela and La Flaca headed off to the López house that morning, I searched for a book that would shed light on this saint who wasn't yet a saint and might never become one. A young man working at one of the religious shops told me that Marianito's canonization process had stalled. To reach sainthood you needed strong advocates in the Vatican, but the Colombian nun in Rome who was supposed to be pushing for Marianito was throwing her energy instead behind another local candidate: the late firebrand Monseigneur Builes, who had spent the 1950s demonizing Antioquia's liberals.

An entranced look washed over the clerk's face as he shared his favorite Marianito miracle with me. In 1994, FARC guerillas took over Angostura, robbing a bank and attacking the police station with rockets and grenades. The army drove them out after a few days, but the guerillas lurked close, hoping to retake the town by night. One night a priest dressed in black appeared to them on the steps of the church, extending his hand to signal: *Get out.*

The guerillas never again attempted to take over Angostura. A bomb they'd planted did not explode: another miracle.

The clerk sold me a pocket-sized volume that contained stories of how Marianito, during his lifetime, had exhorted swarms of ants and locusts to leave peoples' crops alone, how he'd caused a plague of rats to drop dead with holy water and incantations, and how he'd healed so many sick people by rubbing them with his saliva that residents of other towns wrote him to request jars of it.

When I knocked on the office door of the church, to see if I could learn more, the parish priest, Rodrigo Cifuentes, invited me for coffee and a tour.

"Everything is Marianito here," said Cifuentes, at a table facing the gorgeous central garden of the rectory Marianito had built. This was good and bad, he said. People came in droves to Mass, but as soon as they entered the church, they made straight for Marianito's remains.

It had been ninety-two years now since Marianito died, Cifuentes pointed out, but people in Angostura still credited Marianito for every good thing that happened to them. Of the countless miracles attributed to Marianito, most were modest. Cifuentes gave a recent example: A woman had come to him asking to be blessed in Marianito's name. She wanted to buy a car, but didn't have the funds. Shortly thereafter she was able to purchase the car—thanks, she said, to the intercession of Marianito. Cifuentes had to respond carefully to these stories, so as not to diminish people's experiences of faith, but they didn't help Marianito's case for sainthood.

"Things like that don't count for Rome," the priest told me. "Rome wants to see medical miracles, because they can be certified by a doctor."

I had been combing the Marianito book for any reference, any clue, suggesting that Marianito had cured someone of dementia, whether during his life or in his busy afterlife. Generations of E280A family members had begged for his help for their loved ones with Alzheimer's. Some of them were even Marianito's kin, descendants of his siblings.

I found nothing. Cifuentes, who had lived in Angostura for five years, had heard of no such case either. He'd gotten to know many of the E280A families, knew about Lopera and his studies, and had a good understanding of the disease's course. Depressed, overwhelmed caretakers of patients sought Cifuentes's help, while young people often asked Cifuentes whether they should marry someone from the affected families, worried about what the future might bring. "But in the end, they always do," he said.

DANIELA AND LA FLACA, after a day spent swimming and kicking around soccer balls with their López cousins, were waiting for me outside the rectory. They'd located their other cousins, the Pinedas, they said: the family owned a cosmetics store on the plaza. The store was run by two Pineda sisters.

Now all Flaca and Daniela had to do was screw up the courage to greet them. They poked their heads into the doorway of the store, where two dour-looking white women sat behind the counter.

"Are you Sol and Mariana Pineda?" Daniela asked.

"No," they both said, shaking their heads.

"You're not?" asked Daniela.

"No," they insisted blithely.

Daniela and La Flaca nodded and backed out of the store, reeling.

Neither knew how to respond. "I mean, it's one thing if they didn't want to talk to us, or they weren't interested," Daniela said as we walked away. "But to deny they're even related?"

She stopped a man in the middle of the plaza—the same man who had guided us to the old López home days before. She asked him who owned the store. The sisters, I saw, were watching us from inside.

The Pinedas, the man said.

And who were the women inside it now?

The Pinedas, he said.

La Flaca winced. "Disgusting," she said.

Shocked by the encounter, we retreated to the hotel. As Daniela and Flaca returned to their room, clucking with indignation, I stayed in the lobby, nursing a cup of coffee as I chatted with the hotel's owners. When I told them what had happened—that two of the Pineda sisters had humiliated their cousins by denying their identities—the couple didn't seem surprised.

The sisters were weird, the man said: "It's because that family has an illness called Alzheimer's." He said that their brother, Diego, was now ill; this was the man who had married Uncle Javier's daughter, I realized, and had a child with her.

Diego had been one of his best friends, the man said, yet the sisters weren't telling anyone where they could visit him. No one knew where he was living.

"They're pleasant enough when you're a customer," his wife said of the sisters, "but they're known for being standoffish and secretive. Which is pretty rare for a town like this."

SIXTEEN

In the afternoons, starting in late September, storm clouds drag heavy rains and lightning over Medellín, one district at a time. The mornings fill with the honking cries of southern lapwings, stilt-like birds with red eyes that flock to mushy, waterlogged fields. And then one day in early December, as Christmas is getting into gear, the rainy season ends abruptly and the ground dries up. The nights become clear again, just in time for lights to be strung all over the hillsides and speakers set up on patios.

The 2018 Neurociencias Christmas party was less festive than the previous year's. It was just a lunch, with no entertainment, and Lopera's annual address would bring mostly bad news. He began with a litany of the anti-amyloid therapies whose trials had been suspended because the drugs were deemed ineffective or unsafe. Three trials of crenezumab, including API Colombia, were still marching forward without serious safety issues, Lopera reported, but the Colombia trial's switch to a higher dose was still being held up by regulators.

The investigators had grown worried about keeping people in the trial, even though their dropout rate had so far been low. After five years only about a dozen participants had quit, most of them women who had, or wanted to, become pregnant. But there was still a fear that more people would leave. The trial staff acknowledged that some participants now had symptoms of dementia. This was entirely expected, regardless of the drug's effectiveness; half the E280A-positive participants were receiving placebo. Seeing it happen was another matter. No one wanted people to become demoralized and withdraw.

Lopera whipped through his science presentation, a refresher on Alzheimer's disease and amyloid plaques. With that out of the way, he addressed the group about some things that had been bothering him. First, he reminded people that they needed to be on time for their trial appointments, and to stop selling stuff in the waiting room. It wasn't rare for participants to peddle lottery tickets, cosmetics, and clothes; that was how many women made money in Medellín, and a majority of the trial participants were women. Also, he said, he had heard that some of the participants were referring to themselves as "guinea pigs"—or that their families and friends called them such. This clinical trial, he said, was being conducted under the watch of an ethics committee, government regulators, and data safety monitors, he reminded them, sounding peeved. No one in this room was a guinea pig.

He brightened up talking about that year's big research development at Neurociencias: new families with presenilin mutations besides E280A, some of whom had been flown in for this party. Besides the family from Girardota, there was a group from the city of Montería, on Colombia's Caribbean coast. Lopera sounded thrilled to welcome the newcomers to the Neurociencias fold, but the E280A families' response was muted. A research group that until then had focused on a single, homogeneous clan, with whom the investigators shared cultural, geographic, and even family links, had broadened into something bigger and more inclusive—an exciting change for the researchers, but one with uncertain implications for the families.

The Christmas party always served as a forum for the research participants to have their say, to weigh in on how things were going.

Inevitably Lopera faced the same question he always did about genetic disclosure: When will we get to find out? Likely in 2022, he said, without making a firm promise. In any event, he went on, it would involve a formal process, in which participants would undergo psychological screening "to make sure no one commits suicide." Those who didn't want to know, he said, could "go on peacefully with their lives."

Lopera often mentioned suicide when the issue of disclosure came up, although I later learned that no one in the cohort was known to have carried it out. I was not sure why he was still hedging on disclosure at this point, except that the Banner scientists, who handled much of the logistics on the trial, were also unsure about how to approach it.

Eric Reiman and Pierre Tariot hadn't decided when genetic counseling should be rolled out. They didn't know whether to offer it to trial participants only, or to the thousands of E280A family members whose data—whose bodies—had made the trial possible. They didn't know whether to do it immediately after participants received their last study doses of crenezumab, or after definitive trial results were in, or at some later point. And they still didn't know how it would be paid for; some source of funding had to be found. "At this moment, I can't look anybody in the eye and say, 'Here's exactly what's going to happen,'" Reiman had told me.

FOR REASONS NO ONE could explain, the Christmas season always saw more brains. They came in from the villages and from Medellín. They came in on Christmas Eve and on New Year's Eve. Every time, the WhatsApp messages flew around and a team of students was assembled, not without difficulty at this time of year: **Who's sober?** one chain began.

The bodies all arrived to the same funeral home, where the lumbering prosector, Muñeton, was often the first on the scene, smoking a cigarette outside. The deceased were young, rarely far into their fifties, skinny and diapered and wearing the religious charms that their families had strung around their wrists or necks. Their clinical histories were riddled with familiar leitmotifs: Getting mad over nothing. Losing tools and words. Unable to wash a plate at forty-seven. Not talking by forty-eight. Two years in bed, suffering seizures and finally sepsis. The Neurociencias team worked to the sounds of the morticians' radio, which was now playing "Feliz Navidad" and the jumpy party tunes that repeated all month without reprieve.

In previous years most of the donations were overseen by Andrés Villegas, but increasingly the brain bank was managed by the young physician David Aguillón, who had become so indispensable to virtually everything Neurociencias did—the clinical trial, the autopsies, the home visits, the search for new mutations—that it seemed clear he was Lopera's heir apparent. Aguillón was a slight, bearded man in his late twenties, with a reedy voice and an excess of that wasplike energy cherished by the paisas. He had spent his career to date with the group, studying with them as an undergraduate, working with them through medical school, and finally joining them as a doctor. He lived in a rented apartment just blocks from the SIU, in a dangerous part of town, so that he could show up to work on a moment's notice. Like Lopera he was a talented clinician, having seen more cases of young-onset dementia by now than most neurologists do in their entire careers; he could pick up rare diagnoses in no time. Lopera always said that Aguillón reminded him of himself in his younger, more frenetic days.

Aguillón possessed little of Lopera's country charm and magnetism—he was dry and to the point with people—but he tended to get what he was after. Part of his job was to engage the new families, and he'd just persuaded the Montería clan to join the DIAN studies, handing them consent forms to sign during the Christmas party.

LAS FAMILIAS

The investigators had now amassed evidence of eleven distinct presenilin mutations in Colombia, five of them never reported elsewhere. Most produced classic early-onset Alzheimer's, but at different ages and with some differences in symptoms. Aguillón's work on the clinical characteristics of the disease corresponded to what Ken Kosik and his student Juliana Acosta-Uribe were doing with the genotypes. Kosik and Acosta-Uribe sought to identify each mutation, determine whether it was unique or had been seen elsewhere, and—fresh from the revelation that the Girardota mutation originated in a person of African descent—define its local ancestry.

BY NOW THE NEUROCIENCIAS team had seen nearly one hundred of the Girardota family members in the clinic and were preparing the first paper on their disease. Lopera loved the idea that there were two unique paisa mutations—one European, another African. He'd stopped calling Antioquia a "genetic isolate," as the group had long done. It was evident to him by now that the theories about racial purity and disease in Antioquia were baseless.

I hadn't encountered anyone from the Girardota family at the Christmas party; perhaps they'd been invited but didn't come. Lopera, eager that they should feel part of things, had taken Piedad Rua, the woman who spoke for the family and dealt on their behalf with the researchers, to Chicago with his group that summer. Later, after I came to know Piedad, I marveled that she had ever stepped onto that plane. I'd met her briefly once at Neurociencias, when Randall Bateman from DIAN came to pitch the family on his research group and his drug trial. A jittery, elegant woman in her sixties, with tawny skin and red fingernails, she'd told me a little bit then about her family.

Girardota, where they lived, sits just past the northern limits of Medellín, a quick bus ride from the last stop on the metro. It ascends from the banks of the Medellín River up to the town's central plaza and cathedral, a giant brick structure housing an eighteenth-century wooden statue of a fallen Christ tied to a whipping post. During Holy Week, thousands of Medellín residents make pilgrimages to the *Señor caído*, walking the whole twenty miles to Girardota in hope of a miracle, or to give thanks for a past one. The town's central plaza is a happy, social place, with painted murals and cafés and colorful public Jeeps that ferry people to hamlets in the hills across the river. The hamlets are but a stone's throw from the modern and connected town, but home to cane fields and sugar mills and taverns where you can tie up a horse.

I visited on a weekday with the neuropsychologist Sonia Moreno, who sought to add to a family geneaology still only two or three generations deep, with no hint of a common ancestor. Piedad's front porch opened to a splendid view of green hills, interrupted only by the chimney of a paint factory that blew noxious fumes her way. Just below Piedad's house was the cemetery where five of her siblings now rested. Piedad had cared for several of them in the end stages of their disease, and also managed their remains, because in Colombia, unless you were very rich, no resting place was truly final. You interred your loved one in an aboveground rented crypt, waited several years until the body was reduced to bones, cracked the thin cement seal, opened the casket, and removed the bones to a bag. The bones could then either be cremated or go straight into an ossuary, a smaller, more affordable space that could fit several sets of remains. Handling bones could be traumatizing, and it was no wonder so many families now opted for immediate cremation. But in Catholic strongholds like Girardota, people still did things the old-fashioned way. Piedad had recently bagged up the desiccated remains of her sister Trinidad, and for days afterward she cried.

Two more of her siblings were now sick: a brother, Álvaro, whom I'd seen in Neurociencias, and her youngest sister, Camila. That made seven out of ten in her generation: the same as the López Pinedas. Piedad helped care for Camila, alternating shifts with her brother-in-law. Without any full-time patients in her home, Piedad had more time to herself, and more money from her pension, than she'd had in recent years, when her every resource went into the care of others. Yet her anxiety was palpable as she served Moreno and me a delicious chicken stew that she did not eat with us, apologizing that it probably tasted terrible.

For a long time, Piedad explained, she did not have time to eat, and she'd gotten into the habit of not eating, losing so much weight that her clothes fell off. Whenever she answered the phone, she said, she assumed she was about to hear bad news. She had developed a sort of agoraphobia after having to attend to people who couldn't be left alone even for a minute, and she hated to leave her house. She even quit going to Mass, though she had done so her whole life. Over the years many priests and nuns had come to Piedad's home to deliver sacraments to her gravely ill charges. They told her, admiringly, that her relentless caring at the expense of her own well-being was the cross sent to her by Christ. Piedad said she had recently consulted a doctor who prescribed her an antidepressant and a sedative, but she refused to take them. She'd been forced to give sedatives to Trinidad and Camila when they would not sleep, but she had felt guilty drugging her sisters and worried that the pills had further harmed their delicate brains.

Piedad went off to smoke a cigarette in her garden while we shuffled through a pile of documents: ID cards of her dead siblings and parents; cemetery receipts; hospitalization records; copies of baptismal, marriage, and death certificates from the diocese of Girardota.

Piedad's parents had roots in the same surrounding highland hamlets that Piedad's house looked out on. When an earthquake destroyed their farmhouse in 1979, Piedad's parents moved with their children to a plot at the edge of town, leaving rural life behind. Piedad's father, Horacio, displayed the first symptoms of Alzheimer's soon afterward. He ignored food he was served and got lost returning from church. He grew aggressive. Piedad would return from her shift at the factory to help her mother bathe him, holding back tears as he kicked and punched them. Horacio died in 1984 from what his doctors called senile dementia, the same disease that had claimed his father and two of his siblings.

For nearly twenty years after Horacio's death, the family enjoyed a respite from dementia, which they called different things: a person in the early stages was *desmentizado*, a word used locally to mean "out of their mind," while someone in the last stages was *tullido*, crippled. They were uncertain as to what caused it, but Piedad's mother suspected witchcraft. Unlike her siblings Piedad never married or had children, choosing to stay home and support her mother.

In 2002, Piedad's sister Oliva, who lived upstairs with her family, came down in tears. Was it normal, she asked Piedad, to have to ask the kids what day of the week it was? A year or so later, Oliva got lost in a bus station. A local doctor referred the family to Neurociencias. Lopera, Lucía Madrigal, and their colleagues converged on Piedad's house with cots, tests, and syringes, inviting the whole family to take part.

Lopera and his colleagues thought the family's disease had to be early-onset Alzheimer's caused by a presenilin mutation. But when their blood samples came back, they were negative for E280A. Ken Kosik, who had attended one of the visits to Piedad's house, hadn't yet grasped the significance of a possible second mutation in the backyard of E280A. To him, the Girardota family's location, so close to the E280A clan, suggested that the E280A mutation might have duplicated itself in the past, emerging in them as something slightly different.

Oliva died. So did Piedad's brothers Cenen and Alberto and her sister Otilia and finally, Trinidad. The investigators told them that their disease had all the hallmarks of Alzheimer's, though without autopsy brains this could not be confirmed. Their friends and neighbors had other ideas. The usual suspects were blamed: jealous women, witches, malevolent priests. "We were brainwashed into seeing ourselves the way others saw us," Piedad said. "And we were so desperate that we went anywhere anyone told us to go for help, and wasted a lot of money on quacks." Piedad believed in the scientists, but her siblings' families were wary, refusing to donate their brains when they died. "Worms ate them," she said. "What a shame."

PIEDAD, REFLEXIVELY CONTRITE ABOUT everything, apologized for not knowing more about her family's past. But much of it seemed to relate to sugar. Her father, Horacio, had cut sugarcane and played the guitar to accompany the sainete troubadours in the hills. Her uncles worked in the small, rustic sugar mills that dotted the hilltops around Girardota. Two of her cousins died as children by falling onto one mill's machinery, a memory that made her shudder. Piedad had worked all her life manufacturing the bags used to pack sugar.

Moreno and I told her what we had learned about the I414T mutation, that it had originated in someone from West Africa, from what is now the Gambia. No one knew exactly when that person had arrived in Girardota. What we did know was that by 1664, more than one hundred enslaved people of African descent were working in the region's mines, cane fields, and homes. Ten were registered as born in west or west-central Africa—Guinea, Congo, Angola, and Cape Verde were among their places of origin—and had entered Colombia through the slave port of Cartagena.

That afternoon Moreno went fishing in Girardota diocese archives for the ancestors of Indalecio Rua, Piedad's paternal grandfather, who was said to have died of dementia. She retrieved a certificate revealing that Indalecio had been baptized in 1865 as an *hijo natural*—a son born out of wedlock—to a woman named María Mathilde Rua. María Mathilde, born in 1844, was herself an *hija natural*. She was the daughter of one Manuela "Ruda"—a likely mistranscription of Rua—and the granddaughter of a married couple named Pastor Ruda and Paulina Meneces. Pastor and Paulina would have been born around 1800, which meant that one or both might have begun their lives enslaved.

Church records went back no further, and without knowing who the fathers of María Mathilde and Indalecio were and at what age everyone had died, it was difficult to trace with whom the I416T mutation had traveled. Later Mary Roldán, a historian of Colombia, sent me a document listing "Paulina" and "Pastor" among the names of slaves held in 1808 by Manuel Londoño, a wealthy landowner who had commissioned Girardota's famous statue of the fallen Christ. Londoño was a fervent Catholic who encouraged his enslaved couples to marry; upon his death he had willed some of those couples and their children freed. As in the American South, the norm in Colombia was that landowners' surnames got imposed on their slaves. But why would Paulina and Pastor, if they were born on Londoño's estates, have ended up with other surnames? When I asked Roldán, she said many people changed their names after slavery.

Beyond Pastor and Paulina stood another century and a half of unknowns. To get all the way back to a ship from West Africa would require consulting the kind of records that even historians had difficulty obtaining, and the chances of success were slim.

"LET ME UNDERSTAND SOMETHING," Piedad said as we wrapped up. "These illnesses that we have, that we suffer in Colombia, were brought by the Spaniards and the slaves that arrived so long ago?"

"It does seem that way," Moreno said.

So why, Piedad demanded, hadn't the effects of this disease diminished with time? It seemed unfair that her nieces and nephews should have to suffer the way her sisters and brothers had. The next generation of the Girardota family was upwardly mobile—more so, I would come to think, than the young adults in the E280A families, whose parents had experienced a more abrupt transition from rural to urban life.

Piedad had some 150 nieces, nephews, and second cousins, most of them still too young to become sick. In recent years, when Neurociencias came calling again, she had all but shoved them into the researchers' arms, hosting meetings in her house and demanding that they participate. It was for them that Piedad had boarded a plane to Chicago, when she could barely bring herself to go grocery shopping in town every other week. Piedad was aware that disease research was a slow process, and that it would not help her own generation much to be studied. "We're nearly seventy, we're practically about to die," she said. "But for those coming after us—if we really love our family—we should take part."

SEVENTEEN

Daniela López Pineda arrived in Medellín brimming with optimism and happy to feel the Antioquia sun on her face. I met her downtown, near where the buses from the airport stop, and she filled me in on what had happened since we'd last seen each other the previous summer.

She'd been planning to abandon her life in Bogotá, where she had taken a job cleaning the offices of a real estate company. She would break it off with Ximena, whose selfishness she was tiring of, live with La Flaca and look for work. But when Daniela went to give the head of her company notice, the woman ended up talking with her for hours. The boss was intrigued to learn that the cheerful, energetic *muchacha* in a uniform and hairnet had an associate's degree in office administration. Would she be interested in interviewing for a job as a bookkeeper? Daniela returned days later in dress slacks and a blouse. It was the first time since her mother's death that she had been offered a job worthy of her two years of college. She could not pass up the opportunity, even if it meant continuing to live with Ximena until she had the means to move out.

Daniela had to cancel her planned escape to Medellín, reducing her move to a weeklong vacation, to her sister's chagrin. She would hang out with Flaca, buy some cheap work outfits straight from Medellín's sweatshops, and visit her mother's remains, which were locked away in an ossuary belonging to her detested stepfather's family. And there was one more thing she hoped to do: learn her genetic status.

Some of the E280A family members had found workarounds to learn whether they carried the mutation, including traveling to Spain or the United States to be tested. More recently, a handful of people had successfully petitioned their insurance companies for a test. But the families were not routinely informed of such options. I remembered Andrés Villegas hearing out Daniela's pleas, then advising that she get tested—without saying how or where. The investigators associated with the clinical trial always made it sound as though geneticists were scarcer than unicorns in Colombia.

As it turned out, there were several in private practice, including in Medellín. Daniela was able to book an appointment within days. She did so without telling her sister, and I accompanied her instead. She was wet-haired, perfumed, and full of nervous excitement when we met outside the hospital where the geneticist worked.

The geneticist, a woman barely older than Daniela, took immediate interest in her case. She began by drawing chromosomes, and she explained autosomal dominance, a concept in which Daniela was already well versed. The geneticist knew how things were done at Neurociencias, and said it was not uncommon in research settings to withhold genetic results. What her hospital offered, she explained, was not research but a clinical service. The results of Daniela's test would be reviewed by the hospital's chief geneticist, who, along with a psychologist and neurologist, would disclose them to Daniela.

Daniela scheduled the blood draw for nearly eight months later. By then, she calculated, she would have saved enough money at her new job—the test and consults together cost in the realm of a thousand dollars—and earned the right to a few days off. I commented that eight months felt far away. "Not if you've been waiting for something all your life," she said.

The visit had been an eye-opening experience. The private hospital was a mere bus ride away from the SIU. Information that thousands of E280A family members had been conditioned to consider morally dangerous and off-limits was attainable without the slightest stigma, as long as you could pay for it. I figured that La Flaca and María Elena and the other López Pinedas would do everything in their power to talk Daniela out of this, and they had eight months to do so, but for now, as we walked in the blinding morning sun, Daniela reveled in her newfound sense of possibility. "I want my result for me," she said. "I'm not telling Flaca or my aunts. Look, I'm twenty-six. The way I see it, if I have the gene, my life plan won't include being a mom. If I end up never knowing, and all this time I never had the gene, I'll have wasted my chance." She described her emotions as "between fear and excitement, between sad and happy. Because obviously I will know the truth. I will know where I stand in this whole genetic game."

We headed to my apartment, which was close to the hospital, to rest a while. That way, Daniela would not return directly to her sister and be tempted to blurt it all out. She was still in my apartment, about to leave, when I received a text from Ken Kosik.

BAD FUCKING NEWS, he wrote. Roche had just made an announcement: it was canceling its two crenezumab trials outside Colombia.

In these trials, researchers had concluded, the antibody was ineffective in people in the first stages of Alzheimer's disease—even at the high, intravenous dose, which the Colombian cohort had yet to start on. Roche had stressed in its public statement that the API Colombia trial would continue even after the other trials closed, but no one knew for how long.

Daniela stared at me from the couch, demanding to know what was going on. The other two crenezumab trials were suspended, I told her. The drug was found not to work. No one in Daniela's immediate family was enrolled in API Colombia. All the same, the implications of the news hit Daniela hard. All morning we'd talked so hopefully about her future. How long would it be before a viable treatment appeared?

When she left, I read an essay Kosik had sent me along with his text. In it, he and his coauthor, the Nobel Prize–winning biologist David Baltimore, warned that setting deadlines for novel dementia therapies was a bad idea, one that "works against developing evidence for unobvious hypotheses." The goal of having drugs by 2025 had effectively wed Alzheimer's investigators, and drug makers, to the amyloid approach. What the field really needed, Kosik and Baltimore argued, was to diversify—to conduct more basic research into the cascade of events that led to brain damage, even if it meant waiting longer for drugs.

THAT AFTERNOON I HEADED to the SIU. Lopera was away at a meeting, but he'd already instructed his communications office to fend off inquiries and avoid giving details if anyone called.

Francisco Piedrahita, the nurse, was sitting alone in the courtyard when I arrived, looking pensive and morose. The Neurociencias staff were upset, he said, and he imagined Lopera was too. At the same time, he was puzzled, he said, because "everyone had to know this won't work in sick people." There was some undetermined moment when the damage became unstoppable, Piedrahita maintained, when drugging the Alzheimer's brain was like "putting a raisin in water and expecting it to turn back to a grape." Piedrahita had seen the startling results of amyloid and tau scans in seemingly healthy mutation carriers—people with heads full of disease proteins who in some cases went on to finish university without incident. When was it too late for the brain to be rescued? All Piedrahita or anyone could hope was that at least some of the API Colombia participants, a group that included several of his own close relatives, had enrolled in time for crenezumab to have some effect.

The next day, when the news of the canceled trials was published in Alzforum, a website tracking research in dementia, investigators from all over weighed in. Some thought crenezumab was no good; its lack of apparent side effects, rather than being a plus, suggested that it wasn't hitting its target. Others shared Piedrahita's view that patients in the canceled trials were already too sick to be helped. Eric Reiman professed disappointment with the failures but said that the Colombia trial would go on. It remained a better test of the amyloid hypothesis, he insisted, than all the failed trials in later stages of the disease.

When Lopera got back from his trip, he didn't seem as rattled by the crenezumab failures as I'd anticipated. "We never thought it would work in sick people," he told me. But Roche, the drug's maker, had clearly hoped otherwise. What if they pulled the plug on API Colombia as well? Lopera's team was not prepared for the possibility of an abrupt halt to the trial. Some forty people would be out of work overnight, and the participants could end up abandoned, confused, and destabilized by the loss of a project that had shaped their lives for six years and counting.

Lopera told me that API would continue. He must have had assurances that it would. And yet, more than a week after the news broke, he had yet to issue a statement on the matter to the families.

DANIELA REGRETTED THAT she hadn't found time to see her sick relatives while she was home, but her sister visited them regularly. After she returned to Bogotá, I went with Flaca to see her aunt Mabilia, who was deteriorating quickly. One day it happens that a person can't walk, Flaca explained. Mabilia, five years into her illness, was now in bed.

When we arrived Mabilia had been moved from the sitting room, where I'd once seen her staring into a mirror, to a bedroom she shared with her twenty-four-year-old son. As before, her nails were painted and there was makeup on her eyes, but she lay in tangled sheets on a futon-style sofa bed that was too low for her to be moved around easily. She was terribly skinny, she hadn't been bathed, and her diapered lower body was exposed. Her mood had entered a persistent fearful state, causing her to tremble and eye her visitors with dread. She repeated one word without pause: "papá"—*papapapapa*—in staccato bursts, calling for Abel, the father who'd died when she was a teenager in Angostura.

Sebastián, Mabilia's son, sat on the other bed looking at his phone, now and then pausing and making clicking noises at his mother, who could not see him. "It's to reassure her, to let her know I'm here," he explained. Mabilia's husband, Checho, was off in another room. He stuck his head in once to offer us a drink, looking and sounding as though he'd been drinking all day. Sebastián, who was Mabilia's only child, had come to blows with his father over her care. Before they'd brought Mabilia's caretaker and sister-in-law, Gloria, into their home, it was Sebastián who bathed his mother, because his father refused to.

Daniela had burst into tears at the sight of Mabilia months before, but Flaca was unfazed, kissing her aunt warmly. She started in with tips and tutorials for Gloria, demonstrating how to roll up blankets under Mabilia's stiff knees, hand her a cloth to grab to keep her from ripping off her clothes, change her position every two hours to prevent bedsores, and clean her mouth with Listerine.

Sebastián listened from his bed as he pretended to look at his phone. Lately Mabilia had been awake all night, a common issue in Alzheimer's, and Sebastián had a construction job he reported to every morning. He'd taken a day off the previous week when a Neurociencias team was supposed to arrive, but they didn't show up until the next day. The hospital bed they promised had yet to come, and the pharmacy had not received the script they were supposed to have sent for sedatives. These researchers were too busy to serve as regular physicians to sick members of the study cohort. But people like Mabilia had no regular physician. Healthcare for the poor was so full of obstacles and delays that it made more sense to many families to just keep calling Neurociencias.

La Flaca smiled as she spoke to Mabilia face to face, close, as to an infant. *Hermosa! Que linda!* The López Pinedas were big on looking beautiful and hearing that they were beautiful. Loud salsa music played in another room, and Flaca grabbed Mabilia's arms, moving them as though dancing. "Do you want a drink of aguardiente?" she asked her aunt. "When?" responded Mabilia. It was the only cogent response I saw Mabilia make that day, her latent consciousness breaking through like a sunbeam. An hour later, the women washed Mabilia's hair, placing a bucket at the edge of the sofa bed as they drizzled warm water onto her head and massaged it with suds. Mabilia calmed with their comforting touch, still saying "papá," but softly, as more of a tick than a desperate cry.

Gloria, the caretaker, told me she was in no hurry to see Mabilia die. "It eventually gets to that point where you want them to die for their own good," she said, but Mabilia was not there yet; she still had a few words she could say, and she could feel love and respond to kindness. Flaca agreed, remembering how, after her own mother could no longer speak, she still answered questions by blinking. When blinking didn't work, she and Daniela would stick a finger in Doralba's mouth and tell her to bite it, something she could do even in the last weeks of her life.

That day I saw an interesting side of Flaca, as loving, capable, and fearless in the face of an illness she might one day suffer herself. She was about to turn thirty-three, the age at which her mother had first hallucinated an owl. On our cab ride home, she remembered the last night of her mother's life, how after her sister's phone call she proceeded in a stupor to the hospital to find Daniela sobbing while cleaning Doralba's lifeless body. She remembered the nurse telling Daniela to stop, not to bother, but Daniela continuing anyway. They had clutched each other at the agonizing sight of their mother being zipped into a black bag. More than a year had passed since.

IN MARCH, THE pharmaceutical company Biogen announced that it had halted two clinical trials of aducanumab, an anti-amyloid antibody that was considered the frontrunner in the race for a new Alzheimer's treatment. There had been little doubt, based on imaging studies, that aducanumab cleared brain amyloid. But it didn't seem to be alleviating the disease's symptoms. Though the patients in the trials had been carefully selected to include only those in the earliest stages, defenders of the amyloid hypothesis once again argued that the drug had probably been started too late, after the neurodegenerative process was unstoppable. Biogen's stock price dropped, wiping out $17 billion of value in a day. Analysts opined that the company would have to do something bold, like purchase another firm, to recover from the blow.

Shortly after the Biogen news I ran into Silvia Ríos, API Colombia's medical director. Ríos was a Russian Peruvian neurologist who had lived in Colombia and lately worked from Canada. A weekend tango dancer who dressed in pencil skirts and billowing designer blouses, Ríos made for a glamorous presence at Neurociencias, where she checked in for weeks at a stretch with the trial participants and the researchers. Ríos said she was dismayed and demoralized by the trial failures; aducanumab had looked even more promising than crenezumab.

Even if crenezumab did work, Ríos candidly predicted, its effects would likely be modest: a partial slowing of symptoms. All across the world of Alzheimer's research, brain amyloid, long considered the most important biomarker of disease, just wasn't tracking with clinical symptoms. You could see the plaques dissolve just as hoped, but the disease stubbornly progressed.

"It's very difficult to develop research without a specific hypothesis," Ríos said. "We neurologists were taught this way. And sometimes we forget there can be other things happening, parallel processes in the brain or in other parts of the body which can trigger or worsen the problem." It wasn't that the amyloid hypothesis was necessarily wrong, she said—there did appear to be some relationship between amyloid and disease—but it didn't explain everything.

Ríos had spent the early 2000s with Neurociencias, visiting Alzheimer's patients in their Medellín homes. "We went to all these dangerous parts of the city and I was shocked," she recalled. "People were so poor and so desperate and living in such dangerous areas. I thought my God, these people, apart from getting this awful disease, they didn't have access to healthcare—nothing." Most of her patients from that time were dead, she said. "No, all of them," she corrected herself.

BY LATE APRIL, three months after the news about crenezumab, Lopera still hadn't sent any word to the patients about the other trials being canceled. Nor had the higher dose regimen been initiated.

I spent a morning with Piedad Rua, who traveled into Medellín for one reason only: diapers. Her monthly ritual involved getting up at

4:00 a.m., taking a bus and the metro, and arriving at a dingy shopping mall downtown by 6:00, all to pick up a prescription of 10 cans of Ensure and 120 diapers for her dying sister, Camila. On the third floor of the mall, in a series of dispensaries for beneficiaries of state-subsidized insurance plans, security guards watched over long queues of poor people, yelling whenever they fell out of form.

Neurociencias doled out diapers to families, but for only a few weeks at a stretch, to tide them over until the pharmacies began dispensing the prescriptions for them. Piedad and I were looking at a wait of between four and six hours. The waiting room was crowded, without seats for everyone. Most of the customers had a beaten-down look, the effects of hardship and perhaps hard living. Like Piedad, many were here to pick up diapers, and a man kept rolling in giant packs of them on a hand truck. "That brand is horrible," Piedad told me as he passed. "It leaks, the urine gets all over the bed." Piedad's ticket said 76, which she considered a good number.

By 10:00 a.m., we had been waiting nearly three hours. I asked Piedad why she didn't ask to have diapers delivered. "You can," she said. "But they never come." We passed time looking at pictures on her phone, where she kept images of her sick siblings. She showed me one of Camila with wild gray hair and a dopey smile, bearing no resemblance to the alluring, serious photo on her government ID.

A team led by David Aguillón had recently checked in on Camila in Girardota. Camila was in the end stages now. Neurociencias lacked any brain from this family, even as an important research paper on them was pending, and Camila's would likely be the first. She was doing badly, Piedad told me, convulsing all the time. Camila's children had little interest in caring for their mother, and her husband had another lover. All of them seemed to want Camila drugged, and they asked Piedad to procure them sedatives.

LAS FAMILIAS

The clerks were calling number 69. It was 10:45 a.m.

"I want to think this all dies with Camila," Piedad said of the illness. But her brother Álvaro was also sick, I reminded her, and there were all the children whose mutation status was unknown.

"I said I *want* to think it's all over. I don't want to think about this mutation anymore."

It was past 11:00 a.m. when Piedad was called, and the clerk had bad news: they were out of diapers. Camila had only five days' worth at home. Piedad would have to come back in a day or two, when she was promised she would get priority in the line. I felt sick and indignant as we left with just the cans of Ensure, but Piedad took it with resignation. We made our way toward the metro, hugging the old stone walls of the basilica and passing the vendors of porn and religious candles and rat poison. We parted ways on the metro steps, where I sat waiting for Flaca, whom I was seeing at noon.

"UGH," FLACA SAID when I told her how I'd spent my morning. She'd waited eternities for diapers when her mother was ill. I said it seemed like everyone in that pharmacy was waiting for diapers. "Everyone," she agreed. But some had unnecessary prescriptions for them, she explained, and were planning to resell them. This was partly why the pharmacy was so stingy. Pirated diapers seemed like a better deal than waiting all day, but with a street value of six dollars a pack, most people chose to wait, in a vicious circle of poverty and desperation and distrust.

Diapers were one of the things the E280A families had specifically demanded when the researchers first canvassed them about a trial. Carole Ho, the scientist who was working for the pharmaceutical sponsor then, remembered diapers as having been a "big issue," but that the company was cautious about distributing goods to clinical trial participants. "You can be viewed as coercing patients by giving them freebies," she explained.

FLACA HAD BROUGHT ALONG her ten-year-old son, and we all walked to a bus stop to meet her aunt María Elena. We were off to see Elcy, the López Pineda aunt who had been living in a government nursing home for six years. María Elena arrived looking chic as always, in a fitted jean jacket with a cascade of brown curls down her back, and we sat together on the ride to the nursing home.

Elcy's nursing home was in the far south of the city, just past the turnoff to Lopera's finca. I pointed out his hill, high and green in the distance, a world apart from the rough neighborhood we were passing through. María Elena marveled at it. How wonderful that he lived there, she said. "I pray for Lopera all the time. I pray that he finds the cure, and that he gets the recognition and honor he deserves in his own lifetime."

María Elena adored Lopera, but she had just made the stark decision to quit his research program. Participating in the Neurociencias studies was making her anxious, she told me, enough so that her family had started to look for symptoms in her. She was tired of it. She'd already told the psychiatrist and nurses that she would not be back for her yearly evaluations, and she would make no more trips to Boston for scans. She was relieved when they responded with kindness and understanding.

Elcy's nursing home stood on the site of an old farm, a generous property with moss-draped trees and patios and gazebos. In an open ward with giant windows we found her in bed, fragile and shriveled, in fresh pajamas and clean blue sheets. Elcy was fifty-six, with the loose skin and sunken jaw of a woman in her eighties. She could no longer speak, but her eyes were bright, and her dark curly hair had been shorn close.

Elcy had worked most of her life as a live-in maid. She had no children. Like Daniela, she was a lesbian, and her longtime partner, with whom she stayed on weekends, sold heavy quantities of marijuana. One day police raided that home and Elcy ended up in jail. Her symptoms started during her incarceration, and after her release, Elcy went to live with her brother Jaime, who wanted to be free of her as soon as possible. As Elcy sank further into disability, María Elena launched a heroic effort to get her sister admitted to the nursing home, pleading with the mayor's office to declare Elcy a ward of the state.

Elcy had survived three siblings who'd fallen ill later than she did. And yet here she was, very much alive, drinking greedily from the juice boxes her sisters had brought her, making little quacks of joy in response to her nephew's kisses. She was in better physical condition than Mabilia, María Elena noted. No one wanted to put their loved one in a home, but the homes knew how to care for someone with Alzheimer's, how to keep them clean and fed.

Flaca hovered at the rails of Elcy's bed, cooing and petting as her son ate a packed lunch alone on the veranda and María Elena stared out at the rolling lawn. It was a beautiful, mild afternoon. María Elena told me she did not take photos of her sick sisters because she felt it was unfair that they should be remembered this way. She resented her mother for having pegged them with this horrible gene. If she ever found herself in this state, she said, glancing back toward Elcy, she hoped that someone would euthanize her.

Did María Elena think she was getting sick? I could not tell.

I had recently asked Lopera how and when Neurociencias delivered the news to someone that they were sick, and his answer was surprising. There was no established protocol. It was all case by case. Often the conversation happened when an ailing patient, or the patient's family, asked doctors to confirm their fears. "If the person comes to the consult saying he thinks he's picado, and he wants me to confirm it, then that's the time," he said. In other cases, a doctor might act to counter the patient's or family's anxious efforts to cover up the disease, denying what was happening. Deciding just when to deliver the news "is a bit of an art," Lopera said. Unless there was a need to deal with a patient's workplace, or start paperwork for a pension, they usually waited until the disease process was well under way.

Lopera's approach seemed like a holdover from an earlier era when patients with cancer and other terminal diseases weren't always informed of their conditions. In recent years, American guidelines to doctors stressed transparency in disclosing a dementia diagnosis, as withholding one denied people the chance to make plans. But in the United States, things like advance directives, wills, and long-term care insurance were also common. Colombian patients had to rely on the goodwill of their families. María Elena did not wish to end up like her sisters. She had just said out loud that she would rather someone killed her. But if she was headed in that direction, she wasn't doing anything about it, perhaps because there was so little she could do.

NOT UNTIL MAY 2019, a full four months after the news of the canceled crenezumab trials, did Lopera finally send his communiqué to the API Colombia families. In it he seemed to be arguing for participants to stick the trial out. Familial Alzheimer's was different from the sporadic, late-onset disease in the other trials, Lopera wrote, and the patients in the canceled trials already had memory loss when they enrolled. His four-page letter seemed to be met with a collective shrug—perhaps its intended effect—and no one dropped out in response.

Kosik showed up that month in Medellín, with a full schedule as always. The most important item on his agenda was straightening out some mess with the brain bank. Until then the only sign of anything amiss had been the intermittent panicked WhatsApp messages about one malfunctioning freezer or another; freezer failures were the bane of tissue banks worldwide. But this was something different. Kosik's group had recently received a million dollars from the National Institutes of Health to compare findings from one hundred of the E280A brains with those of people who had died with late-onset disease, and millions more in funding hung in the balance. As it turned out, though, many of the brains weren't correctly preserved. The problem had gone unnoticed because they'd sat largely unstudied. Only Diego Sepulveda-Falla, the Colombian researcher in Germany, was working extensively with them.

Kosik's studies aimed to use single-cell RNA sequencing to tease out cellular-level differences in gene expression between older people with Alzheimer's disease and people with the E280A mutation. With this technology, he explained, "in every single brain cell you can see what genes are being turned on and off," allowing researchers to understand whether the E280A mutation causes cells to do something different. He hoped that the findings would further elucidate whether Alzheimer's caused by presenilin mutations remained a valid model for the disease in older people. This was hardly a question of mere academic interest. The costly, time-consuming prevention trials in mutation carriers, including the DIAN trials and API Colombia, were based on the premise that what worked against early-onset disease ought to work against late-onset forms.

When autopsy brains are first harvested, Kosik explained, "they should be fixed first in formalin, and then after ten days or so, put in a sucrose-containing mixture with very reduced formalin," he said. For years, however, the E280A brain halves that weren't frozen had been kept in a high-formalin solution, which causes proteins in the tissue to bond over time and makes it harder for antibodies to tell them apart. The frozen halves of the brains were fine. But the group also needed high-quality fixed tissue, and needed it fast.

Andrés Villegas, the brain bank's leader, didn't take well to pressure from Kosik. In January he and Sepulveda-Falla had published a letter in *The Lancet Neurology* announcing a new Colombian–German collaboration to study the E280A brains. The piece began with a provocative claim: because "isolationist policies in some countries" threatened the workings of international science, Villegas and Sepulveda-Falla argued, the Neurociencias brain bank needed new partners. The researchers seemed to be implying that the U.S. government, under President Donald Trump, could no longer be relied on for support. It was true that Trump had proposed cuts to the NIH, but the letter's agenda seemed more a maneuver against Kosik, a signal to the international research community that Kosik didn't speak for the Neurociencias brains. The tussles between the German and American teams had become a headache for Lopera, who was now, five months later, holding a summit.

For a full morning Lopera listened with crossed arms and a serious expression as Kosik, Villegas, and Sepulveda-Falla, who had flown in from Germany, took turns attempting to sell him on their competing projects, making offers and demands. Sepulveda-Falla's work was aimed at determining whether there were molecular differences underlying the varying clinical presentations of E280A cases. Both he and Kosik wanted to connect the brains to their own networks of pathologists and other scientists, and both acknowledged that there needed to be changes in how things were done. Neurociencias required a host of machines: one that automated the antibody staining process, as well as a scanner and a large data server for the images the outside pathologists would analyze. Expert lab techs would have to be hired, and any fresh brains to come in would have to be treated according to costly, unfamiliar, and time-consuming protocols.

Arguing for his plans, Kosik pledged that he would pay for the machines and supplies, underscoring the awkward fact that, at present, only he had the funds. Lopera looked impressed, but Kosik was trampling on Sepulveda-Falla's and Villegas's sensibilities. An American brain pathologist working with Kosik had joined the meeting by video call, and when Kosik casually promised to send him off a Colombian brain that week, Sepulveda-Falla and Villegas balked furiously, then kicked everyone out of the meeting but Kosik, Lopera, and themselves.

I remembered the way Villegas had once talked about the first E280A brain that had gone to Harvard: like a stolen Colombian artifact awaiting repatriation. He and Sepulveda-Falla had reason to be mistrustful. The story of collaborations between scientists in high- and low-income countries was largely one of stolen credit, resource imbalances, and training disparities, not to mention outright sample-grubbing. The Neurociencias team had fielded years' worth of emails from people wanting to get their hands on the Colombian brains but offering nothing in return.

Kosik had spent thirty years trying to prove himself a different kind of gringo. He had just months earlier confronted the Banner scientists over Lopera's authorship on a paper. To Villegas and Sepulveda-Falla, however, he was still a gringo. Sepulveda-Falla, meanwhile, was a Colombian installed at a foreign institution, with no plans to come back. Did a diaspora scientist have a greater moral claim to data and tissues from his country of origin? It was a question I would mull over frequently in the years to follow, as Lopera's former students gained influence from their perches abroad.

"I put Colombia first, always," Sepulveda-Falla replied when I asked him that question. He never just lent his German colleagues the Colombian samples from his lab, he said. He forced them to petition the University of Antioquia—a laborious process that usually resulted in the Germans losing interest. In Sepulveda-Falla's view it was better to have brains that were under-studied, or even badly preserved, than to capitulate to scientific imperialism.

For the rest of that long morning at the SIU, the arguments over the future of the brain bank continued behind closed doors. Finally Kosik, Lopera, Sepulveda-Falla, and Villegas emerged from the conference room, all looking miserable.

LOPERA TOOK ADVANTAGE OF having his collaborators in town to do something he had resisted doing for decades. He convened a group of colleagues to discuss the implications of lifting his longtime ban on genetic disclosure.

The issue had first been forced on him by Randall Bateman of Washington University, who had no intention of budging on the question. If DIAN was to bring its clinical trials to Colombia, Lopera would have to agree to make genetic disclosure available.

In recent months, Lopera had started to reflect openly on the issue. He told me the story of a thirty-eight-year-old woman who'd come to him crying. She was weighing whether to have a baby with her new partner, but she did not know whether she carried the mutation, which would put her at risk of being struck with a catastrophic illness while caring for a young child. "And that made me realize the importance of resolving this situation," he said. "It's time." I later learned from his staff of a woman that age—I didn't know if it was the same person—who had gone on to have children after seeking Lopera's advice. She now had young kids, and dementia. This wasn't a hypothetical situation. This was something that was really happening to people.

But introducing disclosure carried its own risks. For people with adult-onset disorders, counseling was typically reserved for those eighteen and older seeking their own genetic information. You could not seek information on a spouse, sibling, or child. And yet even one's private information had implications for the whole family. If a young adult sought his or her genetic status before his parent became sick, a positive result would reveal the parent as a carrier, with repercussions for that person's well-being. Families could abuse, pressure, or ostracize members with mutations, and fault them for having children. Denying people their status avoided all that.

Lopera had always thought in terms of families, not individuals. He had long demanded, for example, that journalists never reveal the surnames of E280A family members they interviewed, for fear of causing harm to their relatives. When Lopera had something to tell the families, he addressed them in groups. Neurociencias staff conveyed genetic risks in generalized terms: all children of a parent who got sick were at a 50 percent risk for the disease, they said. This was misleading. All children of carriers stood at 50 percent risk of inheriting the mutation, it was true. But each individual person was born with E280A or without it. You had either zero risk of developing the disease, or 100 percent risk.

To introduce disclosure, in other words, gave preference to the individual in a culture where family came first. And this would require treading carefully.

Lopera had called this meeting in part to help clarify what genetic counseling was supposed to achieve. There were no drugs yet that could prevent Alzheimer's, or stop it from progressing. Was the point to support people's personal autonomy, or to discourage them from having babies? The latter idea made him deeply uncomfortable and had caused him tension over the years with his longtime collaborators in the university's genetics department, who now surrounded him at the conference table, waiting to weigh in.

The geneticist Gabriel Bedoya sat down wearing a beret and ready for battle. Bedoya, who had collaborated with Neurociencias for years on different studies, made no bones about that fact that he hoped the E280A families would have fewer children. What had transpired over the past thirty years at Neurociencias, where no one got counseling and virtually everyone had babies, amounted to an "attack on the population," Bedoya declared, triggering Lopera's ire. "You want to tell them not to have kids?" Lopera snapped. "It's not ethical to tell people not to have children. Yes, they need to know they have a fifty percent chance of passing on the mutation to their children, but will we ever tell them not to have children? No, never! Never!"

Bedoya was nearly eighty years old, but his wardrobe of hats, rock T-shirts, and leather bracelets gave him a more youthful aura than Lopera. He was not backing down. "It's more ethical to say to someone who is still healthy, 'You are a carrier for E280A and have a fifty percent chance of having an affected child,'" he griped, than to express risk in group terms. "And if you don't do that we'll be back to the same."

The meeting devolved into yelling. "When I ask people if they'd prefer to be born with the mutation or not born at all, they always say: 'Give me forty-four good years,'" Lopera bellowed. But Bedoya and the other geneticists resisted. This wasn't just about the carriers, they argued, but their spouses, children, and caretakers. How do you reduce the magnitude of suffering without reducing the size of the families?

Just then the word *sterilization* came up, and Carlos Restrepo, a physician and genetic counselor from Bogotá who had flown in for the meeting, interrupted to urge calm. There was no reason for counselors to coerce reproductive decisions, much less sterilization, he said. What was needed, and urgently, was to learn the attitudes of the family members themselves. A carefully designed survey would help to pin down their reasons for wanting or not wanting to know if they were carriers.

Lopera revealed that this was already in the works. With the help of DIAN, Randall Bateman's group, Neurociencias would conduct a pilot program with the Girardota family, which, helpfully, was much smaller than the E280A clan. The first step was a survey to determine whether and why its members desired genetic counseling. Those who said they wanted it would be referred to a local practice; many, presumably, would go on to be tested and learn their results. If the pilot program proved successful, a subgroup of the E280A families—the API Colombia participants—would be offered the same service when the trial ended. Lopera still had plenty of misgivings about disclosure. But he knew, he said, that the current way of doing things could not hold.

It was the first time I'd heard Lopera commit unequivocally to genetic disclosure. His colleagues took turns congratulating him as relief, even joy, spread around the room.

"Bravo!" shouted Gabriel Bedoya, who then asked if he could share an anecdote.

"Thirty years ago," he said, "in the beginning of this whole E280A affair, a young woman showed up with her father who was sick with Alzheimer's, and she said to the folks at the desk: 'I'm bringing in my father to get the gene for Alzheimer's.'" This was how naïve people were then, Bedoya said, how little they understood of the disease or the science. "Now we have people who say: 'My father had Alzheimer's and I want to know if I have the mutation.' This is not the same population that we started with all those years ago. What we need to know is how *these* people, the people we're seeing right now, are thinking."

I thought of Daniela. The young adults I had met at Neurociencias were not their parents and grandparents. They had changed, Colombia had changed, Lopera had changed, and scientific norms had changed since this research program began in the 1980s. Two years later, when Bedoya died from a virus that none of us yet knew existed and was ferried up the corridor of death in a 1930s Cadillac hearse as his students hugged and cried, I would remember him arguing for the rights of the families.

EIGHTEEN

The API trial was approaching its sixth year when the E280A family members finally began receiving their high-dose infusions of crenezumab. It was the summer of 2019, and the Colombian regulators' long delay in approving the study design change had left the researchers anxious. If these more potent infusions came too late in the trial, the investigators knew, the effects could be harder to see. At least the transition went smoothly, and no serious adverse reactions accompanied the dose change.

Before the change, the participants visited the SIU twice a month, receiving quick shots followed by a fifteen-minute wait. Now they came to the newly built trial center, a block away from the old one, just once a month, reclining in infusion chairs as they watched TV, chatted, or napped. During these longer visits, which could also involve medical and psychological checks, they spent hours passing between the clinical area and the lounge, drinking cappuccino from a machine and connecting with their relatives, many of whom they'd barely known before the trial. It was in the trial lounge that I met the children, and sometimes grandchildren, of long-dead patients who had been so important to Neurociencias in the early days.

Among them was a daughter of Lopera's beloved patient Ofelia, who had led investigators to the C2 family of Angostura and who was among the first donors to the brain bank, in 1996. Saida, who had giant brown eyes and a spray of freckles across her nose and cheeks, told me about her free-spirited mother and marveled at the course of her own life—that at forty-four she had become just like Ofelia, she believed. She had no Alzheimer's symptoms that she was aware of, but her eldest brother had been dead more than two years. As a line of clear fluid, either crenezumab or placebo, dripped into her vein, Saida told me that she'd once dreamed that Ofelia was sitting on the edge of her bed, holding her head in her lap and stroking it. It was a dream so vivid that she awoke from it bawling, longing ever since to relive it.

I met the descendants of Pedro Julio Pulgarín, the wandering farmer who in 1984 had wandered right out of the San Vicente hospital and into the frenzied streets of downtown Medellín. The physicians who had treated him could not remember how he'd come back or when, but his daughter Blanca Nidia did. An old boyfriend of hers, a cheese peddler who hailed originally from Belmira, had recognized Pedro Julio; alarmed to see him wandering barefoot and delusional, he'd managed to keep the farmer by his side until he reached the family.

Blanca Nidia Pulgarín was a round little woman with curly gray hair and a voice so powerful that it seemed almost a Darwinian adaptation for communicating in her mountainous surroundings. In the city, people just thought she was yelling at them. The Pulgaríns were farmers, and though they had homes in the town of Belmira, most still lived in their ancestral hamlet upland of the town, a steep slice of green that hovered high above a bend in the Cauca River. Until recent years most of the hamlet lacked access to a road, and one entered or left it only on a horse.

Blanca Nidia was sixty-two and healthy. Some trial participants were in their fifties and early sixties, older than the average age of onset for E280A. This made it likely that most were there as controls, although some people in the cohort did get sick late, so it was plausible that some were indeed carriers. Blanca Nidia had already lost siblings to Alzheimer's; two were in the end stages of the disease now. Like the López Pinedas, her family was divided in its attitudes toward the researchers. While Blanca Nidia and one sister participated cheerfully in studies, another sister refused to deal with Neurociencias, insisting that Lopera sold brains to gringos.

It had not occurred to her that she might be a control subject in API Colombia, and that all the trouble she had getting to Medellín from her finca, where her animals and husband suffered in her absence, was strictly in the service of science. Rather, she figured, the drug must be working.

Blanca Nidia first met Lopera during his days as a neurology resident. They'd reconnected in the 1990s, after her father, Pedro Julio, had died and Lopera was looking to rekindle his Alzheimer's studies. Blanca Nidia was in the first trimester of a high-risk pregnancy then; her two previous newborns had died shortly after being delivered by doctors in Belmira. Lopera had her cared for in the San Vicente hospital, keeping close tabs on her until she gave birth to a healthy boy. Blanca Nidia never forgot that kindness.

Blanca Nidia's eldest son had been murdered more than a decade before, when a drug mafia made up of ex-paramilitaries first pushed its way into the hamlet. Blanca Nidia suspected that he had stumbled upon one of their clandestine cocaine labs. He'd said nothing to his parents, but perhaps he'd told someone else, because a week later police destroyed the lab, and he was found dead.

That mafia never left. They colonized the hills around the Pulgarín farmstead, planting more labs and killing other innocents who bumped into them. For nearly fifteen years, Blanca Nidia had been forced to coexist with her son's killers. At dusk every day their cars could be seen on the road that passed the hamlet, big vehicles with tinted windows carrying "very serious and unfriendly people," she said.

Still, unlike many of their neighbors—and countless other rural Colombians in the clutches of such monsters—Blanca Nidia and her husband had never fled. "It's beautiful land," she said of the hamlet. You could grow blackberries and potatoes and *lulo* in its soils. "No one lacks for food. Whatever you plant grows. If you could see our bananas—they're this big—and the plantains are beautiful." Blanca Nidia, like her father, raised dairy cows. She loved walking with them in the pastures, and she wouldn't trade her way of life for anything, she said.

In the hamlet, people used to refer to the disease as "amnesia." Some attributed it to a trove of gold said to be buried in one of its rocky caves. Was there still gold buried there? I asked Blanca Nidia. "Nah," she said—all the guacas, or Indigenous burial sites, had been dug up by the 1970s, when two brothers discovered a guaca so big that they moved to Medellín and bought a house with the proceeds. Since then, nothing. Plenty of times, Blanca Nidia had gone searching for guacas during Holy Week. She did so alone, because tradition said you must be alone. The hills were steep, but she knew her way around them even in the dark. Even so, her quests were fruitless.

Blanca Nidia was so different from the other men and women in the trial center that morning. They dressed in the flashy style of the barrios, in clothes they might have traded with their teenage sons and daughters, and sat pegged to their phones. Blanca Nidia carried only a simple flip phone. She wore a cross around her neck and an array of scapulars: the Virgin of Carmen, Marianito—"He's in Angostura," she explained—and her local saint, El Señor de los Milagros.

She had two years of education, she said, just enough to learn to read and write; in her day, that was all that was offered in the hamlet. Now things were better. Her surviving son had graduated high school, though he'd had to milk twenty cows every morning before studying. I noticed as we walked that Blanca Nidia sometimes bumped into other pedestrians, as though she wasn't used to maneuvering in populated environments.

THERE WERE TIMES, in that busy trial center, when it seemed that all of Neurociencias was consumed with the goal of seeing API Colombia through, and little else. It had been years since Lopera penned one of his long musing essays. Interests that had captivated the group in the past, such as understanding the origins of E280A, seemed to belong to a different era.

One exception was Margarita Giraldo, a neurologist who had worked closely with Lopera conducting clinical assessments of trial participants, and had been with the group more than thirty years. A gently spoken woman with thick gray hair, it was Giraldo who had first pulled for me Pedro Julio Pulgarín's file from the 1980s, whose contents she knew well, even if others had forgotten them. A longtime friend of the Pulgarín family, she hoped to visit their hamlet again, but with the mafia so active in that region, she hesitated.

Giraldo never ceased to be moved by the stories of the E280A families, which she always took the time to listen to and absorb. She told me about one she had just recently gotten to know. They were a group of siblings from a flower-growing town southeast of Medellín who, while an E280A family, were new to the Neurociencias fold. Their mother had Alzheimer's, she said, and two of the daughters had joined the crenezumab trial.

In the late 1980s, Giraldo explained, Colombia's family welfare agency had intervened to remove the children from their mother's care. The children were sent to separate places, and one was never seen again. A year or so ago, an American woman had gotten in touch with the siblings. She was their biological sister, adopted by a couple in the United States, and she was eager to travel to Colombia to reunite with her birth family. The reunion was happening soon.

She doesn't know anything about the disease? I asked.

"I don't think so," Giraldo said.

I MET TWO of the American woman's sisters shortly afterward in the trial center. Marta Pulgarín Balbín was forty-three, lanky, and green-eyed; people called her *la mona*, or blondie. With her was her thirty-nine-year-old sister Marcela, who sat close by, speaking less and examining the split ends of her long black hair. The sisters shared the same mellifluous, slightly melancholy tone of voice. Both were enrolled in API Colombia, and they came into Medellín together every month; first for injections and later for infusions. Marcela never complained about side effects, but Marta did; the drug produced in her a sensation that she could not describe—not quite ticklish, but strange.

Marta suggested that the three of us go somewhere quieter to talk. We ended up huddled around a steel table in a dingy coffee shop.

"I need to explain a few things about my family," Marta said, and went on to talk for more than an hour in that lilting voice of hers, which cracked now and then from sadness.

THEIR MOTHER'S NAME was Mery, Marta said, and she was the source of much of their pain.

Mery was born in Yarumal, where her father had died young of cancer. Displaced from Yarumal by guerillas, Mery's mother and siblings moved around different towns in eastern Antioquia in the 1970s. The tall and shapely Mery began working in the red-light districts of those towns, where workers on the flower farms, who were tough and hard-drinking like miners, came in to spend their money.

Mery had her first wave of children—Carlos, Marta, Luis, Marcela, and Paula—with a man named Víctor, who abandoned her when the children were young. Every Friday, when she went off to work, Mery took leave of the finca she rented, a place without electricity or running water. The kids were left alone to survive, begging for food from surrounding farms and stealing water from a cemetery, until Mery showed up again Monday. Or Tuesday. "She was one of those mothers with no concern at all for her children," said Marta. Despite this, Mery's near-annual pregnancies continued. She went on to have six more children with five more men, never to form a household with any of them. The family's genealogy showed a pattern of lines radiating from Mery, connecting to men known and unknown like the legs of a spider.

By age ten, Marta was feeding and caring for four younger siblings, including a skinny baby with lavish hair whose name was Yenny, and whom Marta struggled to feed. "She never learned to walk," Marta remembered of her sister. "You would sit her down somewhere and she wouldn't move." Yenny died before her second birthday, "completely malnourished," Marta said, and Mery buried her there on the finca without attracting notice.

The family had to move all the time because Mery couldn't make rent. In 1987, when Marta was eleven, Mery plopped the children on a scrap of land, owned by one of her brothers, in a hillside barrio of Medellín. They put up a primitive structure of boards and a plastic roof. Again there was no plumbing or electricity; the kids placed a toilet seat over a creek. Mery's latest pregnancy was causing her to bleed, so she left the children alone while she took bed rest at her mother's in the countryside.

In the late 1980s in Colombia, a time when there were homeless children all over the streets, a family had to be in dire shape to attract the attention of the welfare agency, and Mery's finally did. Neighbors watched Marta and the smaller ones fend pitifully for themselves, waiting for Mery to reappear, until they couldn't stand it anymore and reported them. The officials piled all the kids into cars and, going on the little the children could tell them of her whereabouts, located Mery at her mother's house. A very emotional, very pregnant Mery protested her innocence, but the officials were unmoved. It was clear to them that she had no way of caring for the number of children she had, so most were placed with relatives. Marta and Paula each went off to live with aunts they barely knew. Within a year, Paula had disappeared.

As adolescents and teens, the surviving siblings—minus Paula—made their way back to their mother. Mery was living in the town of La Ceja, a hub of the region's industrial flower farms. There she was cleaning restaurants, taking in laundry, and sometimes selling drugs to feed the younger children she now had. Every year during the festival of the Virgin of Carmen, the parish in La Ceja raffled off a new home to a needy family, and in 1992, Mery won that raffle, moving into a simple rowhouse with all the children she could squeeze in with her. Several of those children would go on to develop serious problems with drugs and alcohol, but their life together in those years settled into that of a normal paisa family. There were birthdays, holidays, special lunches on Sundays. When she turned eighteen, Marta started working on flower farms, packing and cleaning stems. With her spouse, she had three children. Mery attended their baptisms, a doting Catholic grandmother in polyester blouses.

The family knew nothing about Alzheimer's disease. When one of Marta's uncles developed dementia, it was said that he'd been *enyerbado*. When a second uncle lost his memory, it was blamed on the fumes at the tile factory where he worked. But then the aunts—Mery's sisters—started having problems, too. Marta and her siblings did not understand what was happening until some cousins put them in touch with Neurociencias. These were the days of the publicity campaign seeking families with early-onset Alzheimer's, and in 2014, the Neurociencias team took blood, ran tests, and deduced that the disease had arrived by way of Mery's father, who had died of cancer at forty-five—just the age at which he might have begun to develop cognitive symptoms. The investigators were not sure how the family linked to the other E280A clans. But their surnames and origins in Yarumal were suggestive enough. Mery's paternal grandmother, like so many other E280A women of that era, turned out to have died in Medellín's Hospital Mental. This was an E280A family, but a branch that had been missed.

Mery, then in her fifties, was showing signs of memory loss, and the investigators invited her children and their cousins to be screened for the API Colombia trial. Several expressed interest, but they were disqualified over their drug problems. Marta and Marcela were pleased to be chosen, figuring that if they were destined to get sick, crenezumab might help them. "We're forcing ourselves to face reality," Marcela told me. "To try and do something active about this disease."

MARTA HAD YEARNED TO find her lost sister Paula all her life, but had no idea how to go about looking for her, and in the end it was Paula who found Marta, on Facebook.

It happened in 2018, Marta told me, when Paula, in the United States, needed to consult her original adoption papers. She began to contemplate the Colombian double surname she'd had when she was adopted. It hadn't occurred to her before to look for her Colombian family on Facebook, but now she typed those names in, seeking a Marta Pulgarín Balbín. When one appeared in La Ceja, Colombia,

Paula messaged her in her halting childhood Spanish. Marta answered immediately.

God . . . my GOD, Marta wrote when she realized what was happening. **I'm crying.**

I'm crying too, Paula wrote back.

The last time Paula had spoken to Marta was thirty years earlier, during a furtive meeting at a pig farm where their aunts got together with the girls they'd grudgingly taken in. Marta was twelve then, and Paula eight. Paula confessed to her big sister that she was being beaten, and that her aunt held her hands over the stove. What Marta never knew was that soon after that reunion, in October of 1988, the welfare agency—the same agency that had just months earlier placed her with that cruel aunt—responded to a call to find Paula wandering barefoot, malnourished, and wearing tattered clothes. She had not been in school.

Paula was sent to a Catholic orphanage in Medellín. For months the nuns and the welfare agency tried to locate Mery, without success. The next year an American couple arrived to adopt Paula. While still in Colombia, Paula drew a picture of the family she remembered—little stick figures of her siblings and parents—in a notebook provided by her adoptive mother.

After their first exchange on Facebook, Paula sent Marta an image of the drawing, which Marta immediately understood to represent her parents, herself, Paula, and their older brother Carlos. She wasn't sure which infant Mery was supposed to be holding: Yenny or a later baby.

Marta pulled up Paula's profile on Facebook. I saw a dark-haired, athletic woman in her late thirties with a husband and toddler son. They lived on the West Coast, and they seemed to love the outdoors, spending time at beaches and national parks.

Marta had never known where to start looking for Paula. Until

Paula made contact, she had no reason to think her long-lost sister wasn't living in Colombia like the rest of them, working twelve-hour shifts in an underwear factory or on a flower farm. On her trips into Medellín, Marta had never stopped seeking Paula's face in the crowd.

Paula's photos with her friends showed something entirely different: well-dressed, professional women out at restaurants. Marta and Marcela believed Paula to be a doctor, and while her job wasn't clear from her posts, I saw that she and her husband had recently built themselves a spacious house. Looking through more images, I realized that Paula looked a lot like her sister Marcela. But the gulf between their lives was wide. In the dank coffee shop where we sat, the conversation had moved on; in hushed, sad tones, Marcela was talking about a friend of hers from La Ceja whose body had just been pulled from the Medellín River. The friend had been tortured and killed, she related, his eyes and ears eaten by vultures.

SHORTLY AFTERWARD, I WENT to see the sisters again in La Ceja, a busy country town where thousands of flower-farm workers ride bicycles between their homes and the greenhouses. After fifteen years, Marta told me, she'd finally quit the flower trade, to keep the sun and chemicals from aging her further. Now she spent her days attending to her three grandchildren, making drop-offs and pickups from kindergarten on her bicycle. Marcela, who lived around the corner, worked part time and cared for her sweet, chubby two-year-old granddaughter. The two were responsive and tender with the children, all soothing advice and hugs, in contrast to what they'd experienced with their own mother.

With Marta's daughter Yenny, a raven-haired teenager named after her aunt who died in infancy, we formed a group and ambled around town. We walked through a cemetery full of cut roses and hydrangeas arranged by the flower workers to honor their loved ones, passing the freshly sealed crypt of Marcela's murdered friend. As we approached the town center, the sisters pointed out different houses their mother had rented before she'd won her current home. Most were dilapidated or condemned. This district of La Ceja was dark, full of alleys, dead ends, and a pervasive sense of squalor. In the late 1990s, the area had been subject to an extermination campaign by paramilitaries, who used surprise blackouts as cover to massacre local addicts, prostitutes, and drug peddlers. Marcela remembered that as a terrifying time—it was families like hers that were the target of this violent "social cleansing"—but they had survived it.

In the street we bumped into their mother, Mery. She was a tall, slim woman with high cheekbones and cropped hair, and the same pale coloring as Marta. She was dressed smartly in slacks and a vest, walking alongside her sister Elida, who was skinny, shuffling, and disoriented. Mery was sixty-two now, and while she'd been diagnosed with dementia and repeated herself in conversation, she could still cook, bathe, dress herself, and get around. Elida was only fifty-three, yet her impairment was marked; when Marta asked her to tell me her name, it took her three tries to respond.

This was one of the big mysteries of E280A that the investigators were seeking to untangle: why some people resisted the onset of Alzheimer's longer, or progressed more slowly, than others. Elida had no one to care for her now but Mery, who could not be counted on much longer.

We entered the small house where they lived along with two of Mery's sons, both of them addicts who did not work regular jobs. Mery's sitting room was festooned with the customary canvases of saints, the unread Bible on a wooden stand. Mery proved cheerful, with a good sense of humor. Even after all I'd learned about her, it was hard not to like her. She launched into a salty, slightly repetitive tirade about how she'd never managed to cohabit with a man except once, and regretted even that. This was the house Mery had won in a raffle, the house that had allowed her to unite her family after years of desperation.

I kept thinking of Paula, who was supposed to be visiting in just a few months. Marta had been chatting with her for a year, in bits and spurts. Marta grew sad during the long periods when Paula failed to respond—her life in the States was busy—but dearly looked forward to being reunited with her in the fall, when Paula had promised to come. I wondered what Paula would make of all this: the house, the addled sons, the doddering unkempt aunt. Marta had told Paula that Mery had Alzheimer's, and this was part of why she had urged Paula to visit soon, while Mery was still lucid and would remember her. What I wasn't sure of was whether Paula understood that the family's variety of Alzheimer's was aggressive and genetic, and that she, like Marta and the rest of Mery's biological children, was at risk for it as well.

As we sat, Marta mentioned to her mother that Paula would be visiting soon. Mery nodded, paused, then responded with something of a pivot, telling me she'd been working, trying to provide food for her children, when the welfare agency took them away. She'd scraped and scrounged all day for them, she said, even if it meant resorting to begging. Marta shot me a half smile, her eyebrows raised: This was Mery's story and she was sticking to it.

Marta loved her mother despite everything, she told me as we left, and intended to nurse her through her looming illness. "I'm not looking forward to seeing her suffer," she said. It was late afternoon, and Marta and I walked together, just the two of us now, toward the bus station. It was drizzling, and bicycles were starting to stream into town from the flower farms. We stopped in the main plaza and entered the church, where Marta stopped at the sarcophagus of the late monsignor who, in 1992, had handed Mery the keys to her one permanent home. Marta's children sometimes asked her why she wasn't more resentful. "They can't believe the stories about how bad she was," she said of her mother. "But who am I to judge her? I'll leave that to him," she said, pointing to the sky.

I WAS NERVOUS ABOUT contacting Paula, not wanting to put her off her trip. It was jarring enough to be planning a reunion with long-lost family, without dealing with a random journalist who'd inserted herself into their lives. But Paula accepted my friend request.

She told me that she was a nurse practitioner. This profession does not exist in Colombia, so I understood why her sisters had mistaken her for a doctor. She would be visiting Colombia that October with her husband and mother, she confirmed. Though she had heard about Lopera, the crenezumab trial, and the E280A mutation—Marta had told her enough to allow her to look all these up—she didn't seem to want to talk about them. More pressing to her was solving the mysteries that had plagued her since childhood, piecing together the broken chronology of what had happened to her and her siblings. She was eager to be reunited with them, and to visit the orphanage where she'd spent her last nine months in Colombia. I realized that Paula had lived all her life with only the barest clues as to whom she had once been, her head full of loose memories without the connective tissue of known facts.

I don't feel that I have a good grasp of events, she wrote me. **Some of the things Marta said didn't make sense. The reason she told me as to why my biological mom left us didn't make sense to me.**

I told her that her sisters loved Mery and forgave her for everything.

Well, I have a very different view on that, said Paula. **And I know**

that most likely they will be protective of her. I don't mean to be disrespectful, but she isn't my mother. She birthed me but that is all.

They come from a tough place, I said.

I understand that, said Paula. **Believe me. I remember eating out of garbage cans.**

Part 3

Resilience

NINETEEN

The 2019 Alzheimer's Association International Conference, held in Los Angeles, California, opened on a dismal note. An influential health reporter, Sharon Begley, had set the tone in the weeks leading up to it with an article lambasting top Alzheimer's researchers. Begley argued that longtime adherents of the amyloid hypothesis—the "amyloid cabal," as she called them—had "suppressed research on alternative ideas for decades." They influenced what studies got published in top journals, which scientists got funded, who got tenure, and who got speaking slots at reputation-buffing scientific conferences, Begley wrote.

The mounting failures of anti-amyloid drugs only seemed to underscore Begley's argument. Already that year, the trials of crenezumab and aducanumab had been suspended. Then, in July, just days before the conference opened, Eric Reiman and Pierre Tariot reported that Banner Alzheimer's Institute had been forced to halt its two prevention trials in patients at high risk for late-onset Alzheimer's disease. Those trials had used an experimental drug called umibecestat, which stops amyloid from being produced. They had enrolled healthy people in their sixties and seventies with the APOE4 genetic risk variant. But umibecestat actually worsened people's cognition. The Banner scientists described the effect as temporary and reversible, comparing it to drinking a glass of wine. After a few months off the drug, participants would be back to where they were when the study began. Nonetheless, this was a terrible turn of events for Banner, and for a field of research that had been relentlessly centered on amyloid. Things were bad enough that the Alzheimer's Association's chief science officer, Maria Carrillo, addressed the failures in her opening speech at the conference. "It's clear that we need a better understanding of the causes of dementia," Carrillo said, and "new therapeutic targets."

Only Randall Bateman, of DIAN, sounded like he still had a reserve of optimism. "It's premature to call the death of the amyloid hypothesis," he told me. DIAN was now finishing a trial using two different anti-amyloid antibodies, solanezumab and gantenerumab, in people with mutations that caused early-onset disease—a population similar to the Colombian cohort but made up of different families from around the globe. Results were due in several months.

If those results were negative, I asked him, would that be the final blow to the hypothesis?

Bateman didn't think so. Amyloid goes through different molecular forms as it aggregates, he reminded me; it was unclear which if any were toxic or how they affected tau. The amyloid hypothesis "is not about a specific drug, or a specific form of amyloid-beta. It's about the role of all amyloid-beta species in the disease process," he argued. To really test it you'd need to use a drug that suppressed the production of all forms of amyloid, in mutation carriers twenty or more years from their first symptoms.

This notion—of testing anti-amyloid drugs in very young adults—seemed ambitious, expensive, time-consuming, and potentially dangerous, considering how little was known about the risks of long-term use. Years before, Lopera had been concerned about testing drugs on people in their thirties and forties, and crenezumab was chosen because it was considered the safest drug candidate. But Bateman's group now sought to recruit mutation carriers as young as eighteen, for a trial that would use both anti-amyloid and anti-tau agents. Some of those young people would be drawn from the Girardota family.

EVERY YEAR, PARTICIPANTS in DIAN's studies got the chance to get together with the researchers, to share stories and ask questions in a daylong meeting that felt like a more earnest version of Lopera's Christmas party. Lopera attended DIAN's meeting in Los Angeles, bringing some members of the Colombian families along: a woman from the recently identified Alzheimer's clan in Montería, and two people from the E280A families who had won a raffle at the Neurociencias Christmas party.

Bateman began the session with a moment of silence in memory of those who had died of Alzheimer's in the past year. A spate of testimonials followed from members of families that were participating in DIAN's research. Most were young adults who knew themselves to be mutation carriers. One man from Minnesota introduced himself as a "warrior," a term the DIAN researchers embraced to describe their participants. A Canadian woman talked about her decision to freeze her eggs while she took part in trials. A former marine pilot from California, who carried a presenilin mutation traced to Jalisco, Mexico, told the group that he was exercising feverishly, and undergoing a type of brain stimulation, in the hope of warding off the disease.

Globally, DIAN participants had an average of about fourteen years of education, five years more than the E280A families. No one was talking to the DIAN families that day about "garbage" in their brains. Instead the brilliant geneticist Alison Goate, who had worked closely with Lopera and Kosik in the 1990s, walked her audience through a sophisticated discussion of polygenic risk scores and transgenic mouse models, to show them how findings from the early-onset families had contributed to current research.

Some of the DIAN participants even used the same vernacular as the scientists, raising their hands to ask questions like: "What effect size will you need to see to justify an open-label extension?" I didn't know whether the group in Los Angeles was typical or if they represented an unusually proactive arm of the DIAN participants. They weren't getting better drugs than the E280A families enrolled in API Colombia, nor would their extensive medical knowledge protect them from the eventual ravages of their mutations. But they were a whole lot better informed.

Moreover, what DIAN was proposing to offer them in its upcoming trials included costly services I'd never heard mentioned in Colombia: egg freezing and preimplantation diagnosis, in which in vitro fertilization is used and only embryos without a mutation are selected. I snuck glances at Lopera's guests. What were they taking from all this? Most of the members of the recently recruited Montería family, I knew, lived on the banks of the Sinú River in northern Colombia. Kosik and Lopera had just been to see them, traveling by river raft to convene in their homes, many of which were constructions of planks, tarps, and corrugated aluminum.

THE NEXT DAY KOSIK gave his talk about his latest genetic discoveries in Colombia. He and the Neurociencias investigators now logged eleven different presenilin mutations, along with mutations that caused other types of dementia besides Alzheimer's. The mutations sat on European, African, and Indigenous genetic backgrounds—the whole history of the Colombian population, all reflected in Alzheimer's genes. The scientists were able to link some of these backgrounds to different disease presentations. Among the many surprises Kosik and Acosta-

Uribe had discovered in the Colombian samples was a mutation that caused motor neuron disease in Europeans, and frontotemporal dementia when sat on an Amerindian haplotype. Ancestry mattered in neurodegenerative disease, and Latin America turned out to be full of rare dementia-causing mutations on different ancestral backgrounds. The DIAN investigators, too, had been looking closely at ancestry as they logged more Alzheimer's mutations in Brazil, Argentina, Puerto Rico, and Mexico.

Kosik drove home right after his talk. He'd been in an unusually cynical mood at the conference, where the sense of failure was pervasive. "Take an antidepressant first," he said when I told him I was going to go look at some posters. No one had ever accused Kosik of being part of any amyloid cabal. He had studied tau for more than thirty years, and only lately, with anti-amyloid drugs faltering, was the rest of the field starting to catch up. He and his colleagues had spent the past year looking into how tau spreads from neuron to neuron, seeking to identify any genes that suppressed its uptake. "Normally if you add tau to the cell culture, the cells just suck it up," he explained. But there were genes that inhibited that process, by affecting certain proteins on the surface of cells. Kosik had published some of those findings the year before, and lately, he said, Eric Reiman had started asking him to collaborate on a project that made use of them. But Kosik still didn't know what the project involved, because Reiman wouldn't tell him. "Eric's got something up his sleeve," he said before leaving.

JOURNALISTS, TOO, BEGAN LEAVING as the week wore on: without exciting news of a drug, it was hard to generate headlines. I ran into Jason Karlawish, the researcher at the University of Pennsylvania who had helped design the API Colombia trial, as he typed up some notes in the near-empty press room. Karlawish, who remained part of the trial's ethics and cultural sensitivities committee, was writing a book that was critical of the Alzheimer's field. He slammed the recent focus on drugs and argued that what families and patients really needed was support, in the form of home health aides, adult day care, and counseling.

Before, I'd known of Karlawish only from a recorded talk of his on Colombia, a place he'd never visited. Karlawish said he perceived the situation in Colombia as similar to the United States in the 1980s, when Alzheimer's disease was still poorly understood. Whether crenezumab worked or not, he said, the trial "is going to raise awareness about Alzheimer's and the genetics of it in a population where awareness was limited and there's a tense stigma."

This was the first time I'd heard this expressed, among all the different ideas about how the trial was supposed to help the E280A families. If a primary benefit of the trial was to raise awareness of Alzheimer's disease, as Karlawish was arguing, it was a redundant one. Not only was Lopera constantly in the media, but his team reached out regularly to medical societies and insurers to help them understand that early-onset dementia was prevalent in Colombia, and that a forty-five-year-old might very well need an open-ended prescription for diapers or a bed in a nursing home. Was awareness of the disease all the families could expect for their years of service to the cause? What they seemed to want most out of the trial was diapers and free access to crenezumab after the trial was over, if it worked.

Karlawish told me it was unlikely that a trial sponsor would give away drugs indefinitely. I later learned from Rachelle Doody, who ran the neurodegenerative diseases program at Roche, that the company provided drugs only during a trial, or during the period known as an open-label extension, in which patients continue receiving the drug while more data is gathered. The API participants, Doody said, would be "able to access this drug until such time as it became commercially available to them in their country." That meant that if crenezumab did get approved in Colombia, the country's medical insurers, and government, would have to buy it. This was a common, and frequently criticized, practice among pharmaceutical firms working in the developing world: testing expensive therapies in poor populations, then passing the costs on to strapped healthcare systems.

One of the key questions that Karlawish's committee had sought to answer in the run-up to the trial was whether Colombia could provide a drug like crenezumab to people if it got approved. "You wouldn't want to do a study in a country where at the end of the study, you make a great discovery and you say, 'Well, look at this. X does Y in this population,' and people in that country look at you and say, 'There's no way we can really substantively implement that intervention,'" Karlawish told me. "Then you would argue that the study was inherently unjust."

The committee had concluded that the Colombian healthcare system did have the capability to deliver crenezumab. But to whom? I thought back to Piedad Rua and her monthly wait for diapers. Colombia had many wealthy citizens, and many clinics that could provide state-of-the-art infusion drugs like crenezumab. But for most people in these Alzheimer's families, this was another world.

THE CONFERENCE WAS in its final days when Lopera and his collaborators presented their latest findings from the E280A families. Lopera's English allowed for only cursory remarks, and he yielded to the scientist Yakeel Quiroz, a former student of his now at Harvard, who gave him a squeeze around the shoulders before launching into her data. "He is my mentor—I love him," she said.

Quiroz and Reiman reported a joint discovery involving a protein called neurofilament light chain, or NfL. The protein builds up in blood when axons, the long part of nerve cells on which electrical impulses travel, are damaged. In late-onset Alzheimer's disease, NfL begins to rise years before symptoms appear. What Quiroz and Reiman had found was that in E280A carriers, levels increased markedly by the age of twenty-two. That meant that six years before amyloid plaques could be detected in the brain—and more than two decades before the earliest memory issues were likely to surface—carriers had more injured neurons than noncarriers. What was causing this silent damage? When did Alzheimer's really begin?

After their presentation, Quiroz and Reiman were inundated with questions. Quiroz, who was in her forties, was soon to be promoted to associate professor at Harvard Medical School. Her history with Lopera had begun in the 1990s, the dark years in Medellín, when she was an undergraduate studying neuropsychology. The day Pablo Escobar was killed, just blocks from where Quiroz and her family lived, she heard helicopters and gunfire, but she didn't think much of it; helicopters and gunfire weren't that unusual at that time. Conducting scientific investigations in the countryside represented a less familiar sort of danger, and Quiroz had to lie to her father so she could travel with Lopera's teams to the hill towns. Like her colleagues, Quiroz ended up face to face with paramilitaries and guerillas, who would barge into hotels asking questions of the researchers.

It was at Neurociencias that Quiroz first developed her interest in brain imaging—at first with electroencephalography, or EEG, which she combined with neuropsychological testing to get a deeper look at the workings of the E280A brain. In the early 2000s, Quiroz's husband, Joseph Arboleda-Velasquez, then a physician-biologist with Neurociencias, went to work for a spell in Kosik's lab at Harvard. Quiroz followed him to Boston, where she started on her master's degree. She was still working on her PhD when Eric Reiman began taking interest in her imaging studies of the E280A families.

The families—then and now—were central to Quiroz's research. In 2012, Quiroz took a postdoctoral fellowship at Harvard, where after two years she began receiving up to six E280A family members a month from Colombia for short- and long-term studies using different types of imaging. Everything known to date about the spread of amyloid and tau in the brains of E280A carriers had been published by Quiroz and her collaborators. Her team secured visas and flights for people like María Elena, Daniela's aunt.

Quiroz and her husband ran labs at separate institutions within the university, but frequently collaborated with each other, with the Banner scientists, and with Lopera. Their relationship with Kosik was testier and more competitive. Quiroz had become, in my view, the most powerful member of the Neurociencias diaspora, installed in the high-tech, high-budget world of a top U.S. research institution but with unfettered access to the large Alzheimer's families of her birth country, and their decades' worth of data. In Colombia, Lopera appeared omnipotent to me, while in this foreign environment his collaborators were the stars; Quiroz's little hug spoke volumes. Before the year was over, everyone in the Alzheimer's world, and many outside it, would know her name.

THE CONFERENCE HAD ALL but ended when Reiman and Tariot convened an evening presentation in a basement meeting room of a hotel next to the conference center. The Banner researchers were giving an update on the status of their trials. This was something they'd previously scheduled, but they must have been dreading it now amid the fresh news of their failed umibecestat trials. The atmosphere in the room was almost funereal, with people huddled quietly at tables and flower arrangements shadowing the cheese cubes.

Reiman addressed a group that included Lopera, Quiroz, and the DIAN scientists. As exhausted as both he and Tariot looked, Reiman mustered his usual confident posture at the podium.

Since 2002, Reiman began, all trials of Alzheimer's drugs had failed. Every one of them. And that, of course, included two trials run by his own research group, just the week before. "What do these findings mean?" Reiman asked. "How do we stay in the game? What does this mean for the amyloid hypothesis, when people are saying 'Stick a fork in it'?"

Reiman's answer to his own question was one I'd heard many times over the past few days: New targets, new approaches. Continue to test the amyloid hypothesis, but differently. Try combination approaches using anti-tau drugs. Try more antisense oligonucleotides, an emerging class of drugs that acted to silence disease-causing genes.

Tariot followed Reiman at the podium. His voice was hoarse, and when he spoke he seemed genuinely pained, making no attempt to mask the deep personal grief that the trial failures must have caused them.

Tariot often reached for metaphors from music or literature when talking about Alzheimer's disease. The first time I met him, a year and a half earlier, he'd likened people's decisions about genetic disclosure to Beowulf braving the monster Grendel's swamp: Why did one go to that dark and uncertain place? What did one want with the information, with one's remaining time on Earth?

Now he was quoting the German composer Richard Strauss. Strauss, he recounted, had instructed a diva not to play the end of one opera as a tragic farewell, but rather with grace and lightness, "one eye wet and the other dry." That, he felt, summed up where the researchers were at that moment, and how they should proceed. They had much to be sad about, but there were glimmers of hope, and a need to forge ahead. API Colombia, he noted, was still going forward. And researchers were reaping a bonanza of data from the Colombian families, with that week's neurofilament light chain findings just one of many examples.

Then, almost in passing, Tariot made the first public mention of something that would change the trajectory of the entire Colombia project. The researchers had discovered a woman, an E280A mutation carrier, who had had no symptoms of Alzheimer's until her early seventies. When she was flown to Boston for scans, her brain had turned out to have "the most massive amyloid burden ever seen," Tariot announced, "and almost no tau." Intriguingly, the same woman also carried two copies of an unusual mutation on the gene APOE, a gene that—depending on the variant—can increase or decrease risk for late-onset Alzheimer's.

"This may open up a whole new investigative path," Tariot said. "There may be an APOE-mediated connection between amyloid pathology and tau pathology."

I had heard about such cases—rare people with E280A who had lived much longer than they were supposed to without symptoms—but most had been discovered posthumously through revisions of old files and samples. This woman had had recent brain scans. She was actively being followed. She had a brain full of amyloid and almost no tau. What did it mean?

Reiman jumped in to say more about the woman, referring to her as "the case." Not only could the case help unlock the mysteries of Alzheimer's disease, he said, but their team was already "developing a therapy based on it."

On that note, the meeting ended.

On the way back to my hotel, I walked alongside Yakeel Quiroz, who said she was taken aback that Reiman and Tariot had even mentioned the woman's case, because it had nothing to do with the purpose of that night's meeting, an update on the API trials. I asked her about Reiman's tossed-off remark that he was developing a drug based on what they'd found. That's right, she said. She, her husband, Lopera, and Reiman had applied for a patent together, as coinventors. I wondered: Was this the project Reiman had been courting Kosik to participate in, without revealing what it was?

When I spoke to Kosik, he said he knew nothing about a patent, but he knew all about the woman; Reiman, Quiroz, and her husband, Arboleda-Velasquez, had already written a paper on the case that was yet to be published—and they had credited Kosik as a coauthor, because it was Kosik who had done the first genetic studies that included the woman's sample. There was little question in Kosik's mind that some gene had helped the woman stave off the effects of E280A; the question was whether it was the same variant that Quiroz and her colleagues were saying it was. Until they could prove it, the idea of a drug seemed premature. It was an intriguing case, Kosik cautioned, but it was still a single case.

Once I was back in Medellín, I went downtown to the SIU to see what more I could learn about the special patient. At the SIU, there was no secrecy around her at all.

The woman, whose name was Aliria, was seventy-six years old and living in Medellín. She was a member of the C2 family from Angostura, a cousin of the Piedrahitas. Until 2015 she had never been seen in person by Neurociencias, and was known to exist only from genealogies. She had two daughters and two sons, twelve grandchildren, and, I saw, a genealogy that was rife with the kinds of weird deaths endemic to her generation of rural paisas: one brother had disappeared after an earthquake; another died from rabies; a sister had died following a "stomach curse"; a brother had been shot dead in a tavern, where he was playing the guitar.

I immediately went to Lopera, asking him to let me talk to Doña Aliria, as she was known. He promised he'd put me in touch in a few months, once the case report was published.

TWENTY

Medellín's international airport sits apart from the city, in the cool, flower-growing highlands of eastern Antioquia. Over the years I had seen many poignant reunions in its arrivals hall, where giant families await members who'd been missed while they tried to make it abroad. A store right in the hall sold all the posters, confetti, balloons, and streamers required for these occasions, which tend to involve screaming and tearful group hugs.

There, Marta and Marcela Pulgarín Balbín gathered to await Paula, their sister who had been lost to them for thirty years. Paula's flight was scheduled to arrive at 8:40 in the morning. The group had assembled by eight. A swollen-eyed Marta was anxiously orchestrating things, in between bouts of crying. Marcela, with her face made up and her hair blown dry, was almost catatonic with anticipation. Early that morning she had posted a message to her lost sister on Facebook: **The day we have all been waiting for has arrived.**

Their mother, Mery, was there, along with their brother Carlos, a somber man who'd been a teen on the day in 1989 when Paula memorialized him with a picture drawn in crayon. Marta's and Marcela's adult children had come with their partners and children, who carried balloons and unfurled a homemade banner that read WELCOME in English.

Mery was thinner than the first time I'd seen her, her eyes less expressive. One of her children was always with her because she could not be left alone anymore. But she remained mobile, and alert enough to know what was happening. At half past eight Marta started crying again, and her oldest daughter held her supportively. "This is bringing back all kinds of terrible memories for her," her daughter told me.

I still couldn't guess how this reunion would play out. Marta and her family didn't have the resources to serve as full-time tour guides over the week and a half of Paula's stay, and they'd made no plans for their guests beyond the first day. Everything would have to be improvised. By then I'd spoken to Paula several times by phone—she had the same lovely voice as her sisters—and helped her reserve hotel rooms, over Marta's objections. "You'll need the privacy," I assured her. Colombian hospitality meant never leaving guests alone.

On Facebook Paula announced her plans to meet her birth family, receiving messages of support from far and wide.

TWO WEEKS BEFORE, I'd met Marta and Marcela at the trial center, waiting out their long morning of infusions, then setting off with them downtown to spend their API stipends. What they got every month wasn't much—the equivalent of sixty dollars each—but their round-trip bus tickets cost only five dollars, leaving them something to play with. Usually they bought toys or art supplies for the grandkids, but that day they shopped for themselves, snapping up two-dollar sunglasses and little decorative elephants from China.

After buying her fill of tchotchkes, Marta picked up underwear for her aunt Elida, who had recently become incontinent. Elida was still living with Mery—the less impaired sister taking care of the more impaired one, until the day when Marta would have to care for both her mother and her aunt, unless she developed dementia herself. Another of Marta's aunts, Oliva, had just died of Alzheimer's at fifty-five. Marta had shown me a picture of her in the end stages, her face drawn, her mouth open and neck distended as if in agony. What would Paula think of her siblings and their world? She had just emailed to ask whether she should leave her diamond rings at home. I told her that was for the best.

AT 8:45, THE first arrivals from Paula's flight began trickling into the lounge. "I see Sean!" Marcela yelled. She'd seen images of Paula's strapping blond husband on Facebook, and he was the first to come out, recording the scene with a GoPro. With him was a white-haired woman in hiking pants and trail sandals: Miriam, Paula's mother. And finally out came Paula, tired and smiling, dressed simply in jeans. Marta rushed to meet her, and they shared a long hug as the family cheered. Paula then hugged her brother Carlos, her sister Marcela, and the kids, who lined up. Everyone wanted a hug. Then, for good measure, Paula and Marta embraced again.

Marta led Paula to Mery, whom Paula hugged briefly, with Miriam close at her side. Mery addressed Miriam tearfully: *Que Dios le pague*, she repeated. May God repay you. Shortly thereafter, Mery swayed briefly, as though about to faint; she fell back into the arms of her children, who, along with Miriam, kept close to her and fussed over her. Paula walked away.

We loaded onto a bus Marta had rented to return us all to the town of La Ceja. On the bus everyone was mostly silent, as if recovering from the emotion at the airport. Paula and Marta sat in the front, Marta smiling as Paula played with the kids.

I knew then that everything was going to be fine.

Paula had more Spanish than she let on: first languages are amazingly durable. Paula's mother, Miriam, turned out to speak passable

Spanish, which she had taken it upon herself to learn back in the eighties, as she prepared to adopt Paula. In the Pacific Coast town where Miriam lived with her husband and son, she told me, many families had adopted children from Colombia. When the three set off to get Paula, in 1989, all they knew of Paula came in a letter from a social worker, translated into English:

> *Paula is an intelligent child who has a good memory who is logical and reasons well. She is mature for her young age, owing to the many sufferings and frustrations she has undergone. She has a great facility for adaptation and socialization. Her feelings of anger and rancor have been repressed. She wants to be adopted and does not want to be returned to her mother because "she would just leave me (throw me away) again." For Paula to reach her full potential, affection and security are indispensable. Therefore, an adoptive placement is recommended.*

Miriam looked the archetype of the liberal outdoorswoman, a septuagenarian who could climb boulders and leave you in the dust; she carried a notebook that she used to record everyone's names and map the family relations. Sean, Paula's husband, was probably a little lost, but he didn't seem to care, and the children were fascinated with him.

In an hour we arrived at Marta's house. A breakfast effort commenced in the kitchen as Miriam began showing the kids pictures she'd taken of sea lions and whales, causing them to gape and squeal. With every conversation, she entered more data into her notebook: names, places, new words in Spanish.

Paula helped her sisters prepare the chocolate, arepas, and eggs with tomato and scallions. The way Paula touched Marta or Marcela's back when she needed to get in and grab something, the way she moved smoothly around the kitchen with them: it was like she'd never left Colombia.

Mery sat to my right, talking nonstop. At the airport, when she'd first greeted Paula, she'd asked her for forgiveness, then quickly blamed Paula's biological father, Víctor, for what had happened to Paula. Now Miriam listened politely as Mery described how Víctor had beaten her. She was so terrified that he would kill her every time he came home, she said, that she didn't know whether to open the door. He'd bitten her face, Mery said, pointing to a spot under her eye. There was no scar, but Miriam nodded as though she saw one. Mery then repeated the story about the bite.

We all ate breakfast together that morning, perching ourselves on any available surface in a house too small to fit us all. Miriam wanted to know more about Mery's disease. This kind of Alzheimer's was genetic, I told her. Each of Mery's children had a 50 percent chance of having inherited it.

Miriam said she was surprised, given such a high risk, that Marta, Marcela, and the rest had gone on to have kids.

"They didn't know about the disease," I said. It was true: at the time they'd had their children, Marta and Marcela really didn't. And the sisters had only four children between them; some of their siblings had had four or six apiece. But I was not sure that knowing would have changed anything. I could understand now why Lopera had never wanted to discourage the families from having children. Kids were their joy, their hope, the source of so much fun. What else in their lives, or anyone's, could compete? Paula had left her little boy at home with her in-laws, but she wished he were here as she sat with her grandnieces and grandnephews. She was so natural with the little ones, tucking herself in among them and showing them the few snapshots she had of herself as a child, skinny and wide-eyed in the plaid uniform of the orphanage.

We went out walking, forming a procession on our way to the hotel. Paula and her older brother Carlos, who had not said much to Paula until now, walked slowly in front of me, talking. Soon they had linked arms. Paula had ignored Mery since the airport, but with her siblings she seemed to share an instant, genuine intimacy. And Miriam looked delighted to witness it.

Paula's brother Carlos was forty-five and worked for a company that made leather goods. He was more reserved than his sisters, and less attentive to Mery. For years after the welfare agency dispersed the siblings, he told me and Paula, he'd lived mostly in the street. His father, Víctor, had rejected him, and his grandparents refused to care for him. He'd looked for Mery when he heard she was in La Ceja, but she wanted nothing to do with him either. He had never rejoined the fold, never lived in the house Mery won in a raffle. It was only when Carlos was in his early twenties, and got into a serious relationship with a woman, that he began to live normally as part of a family.

Carlos, too, had been invited to take part in the Neurociencias studies. But he was not interested. He felt that his family was cursed, irrevocably—and that in some ways Alzheimer's was the least of it. A couple of his aunts had been as bad with their children as Mery was, he said. All his life he had been haunted by the memory of Yenny, his baby sister who had starved.

"I'm just happy that you didn't have to suffer as long as we did," Carlos told Paula as we stopped in front of their hotel. This remark caused Paula to burst into tears. She hugged her brother tightly. It was the first time in that long, exhausting morning that I'd seen her cry.

Paula's group went up to their rooms for a rest. Carlos left for Medellín, Marcela headed off to her latest job, selling sneakers in the mall, and I joined Marta to help return Mery to her house. It was barely afternoon. When we got there, one of Marta's other brothers sat on the stoop, smoking, as Elida, their aunt with advancing Alzheimer's, appeared in the doorway like a disheveled ghost. An odor of urine wafted out of the house; Elida had wet herself again. When Marta realized this, she joked to her aunt that she'd have to spank her, then chased Elida inside, half spanking her playfully. I remembered something Daniela had said when I first met her: that Alzheimer's was humiliating.

Marta was happy. She sat on the stoop beaming, her knees folded in her arms, the sun lighting up her hair as she reflected on the morning, on how well things had gone, on how much she had to look forward to in the coming days with Paula.

THAT MONTH, THE drug company Biogen announced that it had reversed course on aducanumab, the anti-amyloid antibody whose trials it had halted six months earlier. Since then, the company said, it had reexamined the data. While one trial showed a slight benefit among people in the early stages of Alzheimer's disease, slowing it by about 22 percent, a second one did not. And that might have been the end of it—except, the company noted, that one subgroup of people in that failed trial, those who had gotten the higher dose, had done as well as the patients in the successful trial. Based on these results, Biogen had now decided to seek Food and Drug Administration approval for aducanumab.

Shares in Biogen jumped by nearly a third, and health reporters dubbed the drug "Lazarus." It wasn't just aducanumab whose resurrection seemed miraculous. It was the amyloid hypothesis, on which billions of drug development dollars had been staked. After the failures of the two crenezumab trials in January, all eyes had been on aducanumab, including Francisco Lopera's. He had been following the drug's fortunes all year, and he wanted to believe that the newly announced results were meaningful, but he could not shake his doubts. Biogen had been cagey about its data, showing bits and pieces at conferences and never publishing it in a peer-reviewed journal. "I really don't understand," he told me. "All the anti-amyloid drugs are failing, and aducanumab is a success? What's it got that the others don't?"

I wondered if Lopera had secretly given up on crenezumab. His wife, Claramónika Uribe, who conducted social outreach with the E280A families, had recently told me that the researchers and families alike would have to be "psychologically prepared" for a bad trial result. Lopera and the Banner investigators had had little to say publicly about the trial at the Alzheimer's conference that summer; instead they'd drawn attention to a woman whose case challenged the very premise under which it was being conducted. Aliria had more amyloid built up in her brain than anyone in the API Colombia trial, and yet well into her seventies, she was reportedly doing fine.

I WANTED PAULA to come to the trial center before her trip was over, to meet the neurologist Margarita Giraldo, who had taken an interest in her family, and who could tell her more about the study and what her sisters were doing. I felt that Paula should hear the facts from a medical professional, not just Marta and Marcela. Sean, Paula's husband, had said to me that since Paula was raised on a diet heavy in fish, she was probably well protected from Alzheimer's. Her mother, Miriam, had known nothing of the E280A mutation until their first day in Colombia, over breakfast at Marta's. One thing that stood to mislead them about the seriousness of the situation was Mery's age. Mery had gotten sick late for someone with E280A—at sixty-two, much later than her affected brothers or sisters. This made it appear that the risk to Paula, Marta, Marcela, and the rest was decades away, when it might be only a few years.

For days the family had been having a great time, doing everything in unison. Sean cooked a steak dinner; they'd gone to the mall for ice cream, and visited the church and the plaza. Paula was introduced to her poor aunt Elida, her younger half-siblings, and another of her brothers, who had nothing to say to her—just a hug and a "Hey, sis" before he jumped into a car with some strung-out friends. Paula took everything in stride, it seemed, including the conditions in which her extended family lived. But the one thing she did not and would not do was enter Mery's house. All week and for some time afterward, I learned, Paula had been besieged with memories of her girlhood, provoked by conversations with her siblings. She had been struck by a vision of herself arriving at the orphange, eight years old, her earlobes dripping with pus after someone in her aunt's house had pierced them. She remembered throwing up while riding on a bus next to Mery, then being forced to keep close to Mery's legs in a dark, frightening place filled with men.

BECAUSE MARTA WAS SCHEDULED that week for her semiannual "long" day at the trial center, which involved a brain scan and a battery of cognitive tests in addition to the infusion, the family decided to treat this as another group activity. Paula, Miriam, Sean, Marta, Marcela, and Marta's daughter Yenny all hopped the bus to Medellín.

That morning in the trial center, as deliverymen pushed around hand carts stacked with boxes marked EXPERIMENTAL from Roche, Margarita Giraldo came out to greet us, and she invited Paula, Sean, and Miriam into a conference room to talk.

Paula showed Giraldo the image that she'd drawn of her lost family all those years ago, and Giraldo teared up looking at it, almost in disbelief. Paula sat, composed and attentive, as Giraldo explained the history of the research group, the specifics of the mutation and the study, and how Paula's sisters had come to their attention. In a couple of years, she said, all the participants, including Marta and Marcela, would be given the option to know whether they carried E280A.

Paula took this in without really responding. It was a brief meeting, after which we left to explore downtown a few hours while Marta finished up.

I walked with Miriam, who had not set foot in Medellín since she'd come to adopt Paula in 1989. Until this trip, she said, she'd had no desire to return; nor had her husband or their son, both of whom had declined to come. Their experience of Colombia had been alienating. The three of them were forced to hole up for weeks in a hotel, waiting for the day they could take Paula home. The governor of Antioquia had just been killed by the Medellín cartel. The adoption agent who drove them around was too terrified of being ambushed to even stop at traffic lights. As they flew out of the country from Bogotá with Paula, they peered down from the windows of their plane to see tanks circling the airport. Their effort had felt more like a rescue than an adoption.

As we passed through downtown Medellín, I saw that Paula never flinched. Not at the maimed beggars, the fruit carts blocking every walkway, the prostitutes in their tight dresses.

We spent an hour in the art museum, where Miriam, Paula, and Sean explored rooms full of spectacular giant canvases by Fernando Botero, while Marcela and her niece took selfies.

"I think I'd like to be tested," Paula told me on the museum steps.

I asked her what it would change for her. "Nothing," she said. "I just want to know."

It was the last time during her trip that we talked about Alzheimer's.

TWENTY-ONE

The next news I heard about Aliria, the woman who had resisted the ravages of E280A for decades, came on the front page of *The New York Times*. "Why Didn't She Get Alzheimer's? The Answer Could Hold a Key to Fighting the Disease," read the headline. It was November 2019, and Pam Belluck, the health reporter who'd visited Colombia a decade earlier to write about the E280A families and the clinical trial, had been granted exclusive early access to Aliria's case report. This wasn't how it usually worked with scientific papers, but there was nothing normal about this one. It was a short communication in the journal *Nature Medicine*, just ten paragraphs long, yet signed by a whopping forty-three authors, first and last of whom were Joseph Arboleda-Velasquez—Yakeel Quiroz's husband—and Quiroz. Lopera and Eric Reiman were also listed prominently, along with dozens of collaborators, many of whose names I'd never seen before.

The gist of the paper was that Aliria, who was not named in the paper or the *Times* article, possessed, in addition to E280A, two copies of a rare mutation called APOE3 Christchurch, after the New Zealand city where it was discovered. The APOE gene regulates the transport of lipids from cell to cell, and its three main variants—APOE2, APOE3, and APOE4—confer different levels of risk for late-onset Alzheimer's disease. Every person has two copies of APOE, which can appear in any combination. Having two copies of APOE4 increases a person's risk for Alzheimer's dramatically, while two copies of APOE2 reduces it to nearly nil. APOE3, the most common of the variants, was long regarded as neutral. It was long thought that presenilin mutations like E280A were too strong to be much affected by APOE variants, but the evidence was inconclusive.

The APOE Christchurch mutation was known for causing high cholesterol—which Aliria did in fact have—but it was unknown until now that it had a role in Alzheimer's, too. What the researchers seemed to be arguing was that, in Aliria's case, her rare double Christchurch variant had behaved like a supercharged APOE2, staving off the effects of E280A for decades.

The big mystery of Aliria's case was why Quiroz's brain scans had revealed so little tau in her brain, despite such a large volume of amyloid plaques. The researchers argued that Christchurch had acted on a type of cell surface protein that affected how tau spread among neurons. This was why Eric Reiman had been so interested in Kosik's work. It still seemed likely that some link existed between amyloid and tau. But Aliria had shown that this link could be broken. You could have had all the amyloid in the world and still do rather well, as long as the tau in the brain was restricted.

Though neither the scientific paper nor the *Times* article mentioned the patent application that Quiroz had told me about at the Alzheimer's meeting that summer, the crux of the excitement over Aliria was the discovery of another potential therapeutic target in Alzheimer's disease. APOE was not a disease protein, like amyloid or tau, but a gene that seemed to act on multiple disease pathways. The fact that APOE was the key genetic factor in late-onset Alzheimer's made Aliria's discovery all the more intriguing. Reiman, who had studied APOE since the 1990s and was ever more convinced of its importance, speculated to the *Times* that the future of anti-Alzheimer's therapies might lie in finding ways to change or quell the gene's activity in the brain.

NEWS OF THE DISCOVERY spread swiftly all over the world, including in Colombia, where within days Lopera and Quiroz had both given radio interviews and spoken to newspapers about the seventy-six-year-old woman whose body harbored, in the words of the Colombian daily *El Tiempo*, "both the illness and the cure." In his interview with *El Tiempo*, Lopera described Aliria as a woman of "low economic resources," and said he was seeking donations to pay for the nursing care she would eventually require. Aliria did suffer from memory loss now—she had not beaten Alzheimer's altogether, just delayed it by thirty years, he explained. Having contributed "an impressive amount of knowledge to humanity," Lopera said, she deserved to live her illness "with dignity."

In comments on the *El Tiempo* website, some readers reacted skeptically to the news and to Lopera's plea for donations. If this needy woman was so important to science, they wanted to know, why couldn't Lopera's university pay for her care? A number expressed disbelief that the case was real. On the *New York Times* site, meanwhile, a handful of readers likened Aliria to Henrietta Lacks, the Black American woman whose cancer cells were extracted without her or her family's knowledge, only to be cultivated, sold, and used by generations of researchers. There was no way that Aliria and her family had not consented to the studies, as they'd flown to Boston several times for tests. But people were disturbed by the thought of drugs being developed, and profits reaped, from a poor, uneducated woman's body.

The E280A families, as a rule, did not read *The New York Times* or *El Tiempo*. But they listened to the radio and read the local tabloid *Q'hubo*. After *Q'hubo* published its own, simplified account of the Christchurch discovery, the families began calling Neurociencias. La Flaca and her cousin Lina were all abuzz, hounding me for information about how common the Christchurch mutation was—that is, how likely they were to have it themselves. I disappointed them by saying it was rare. To the scientists' knowledge, Aliria was the only person in the E280A cohort with two copies of Christchurch. There were people with just one, but it was unclear whether one copy had any protective effect.

Kosik and his colleagues had, earlier that year, done their own, differently designed study on genetic variants that might speed up or delay onset of disease caused by E280A. They analyzed whole genomes from hundreds of E280A carriers who had gone on to get dementia, trying to identify different genetic variants affecting the age of onset. Aliria's genome was among those evaluated in that study, but Kosik had not hit on Christchurch. In a genome-wide analysis, he told me, "you'll never find Christchurch, because it's not statistical. It doesn't occur in enough people."

The Aliria paper had three weaknesses, Kosik felt. The first, most obvious one was that it described a single case. The second was that if two copies of Christchurch could counter the effects of E280A, one copy in theory should do *something*—at least delay onset of the disease by a few years. Yet the handful of people Kosik had identified with a single copy of Christchurch got sick at the same time as everyone else. The third problem was that there were half a dozen other outliers with E280A, and some of them had resisted disease just as dramatically as Aliria. None of them had APOE3 Christchurch.

Quiroz and Arboleda-Velasquez were aware of these caveats, of course, which was why they were busy seeking and studying more

RESILIENCE

people with E280A and Christchurch. With a larger sample they could determine whether people with only one copy of Christchurch saw any delay in the onset of symptoms. Another urgent task, Quiroz said, was to see what Christchurch did in experiments with human cells and in animals. A couple of labs were working on this now. Finally, there was the problem of the other outliers in the E280A cohort, who had no copies of Christchurch. One man, who had died in his seventies, was just beginning to have dementia, despite having E280A and a brain full of plaques and tau tangles. What had kept his neurons alive and working with all those disease proteins present? There were further cases to analyze, including very old ones. Lopera recalled a man in Yarumal who had tested positive for E280A, yet died at seventy-six of another disease, with no symptoms of dementia. Recently, Lopera and his colleagues had asked Alison Goate's former lab at Washington University to turn over all the stored Colombian blood samples from the 1990s. They were thrilled when this man's sample turned out to be among them, and they could mine his DNA for possible protective variants.

Lopera felt that Kosik's genome-wide association studies weren't suited to the task of finding rare resistance genes in Alzheimer's, and that Quiroz and Arboleda-Velasquez had the aggressive approach required, even if it was inherently biased. Arboleda-Velasquez had once described his gene hunts to me in terms of a feeling, a hunch, rather than something systematic. Kosik conceded that Quiroz and Arboleda-Velasquez might be right, and that the Christchurch finding represented something truly revolutionary.

I had learned from the Neurociencias staff that Aliria was a chatty, charming lady with a devilish sense of humor. She had lived independently until recently, cooking her own meals and running errands. The family lived in Barrio Pablo Escobar, a tough hillside neighborhood whose first homes had been built by Escobar in the 1980s, at the height of his fame, for families that had previously been forced to live around a garbage dump.

I asked Quiroz if Aliria's family knew that she was the famous case. "They know everything," she said. "They are very supportive. They really understand the importance of the research we are doing." Quiroz wasn't done studying Aliria, and she expected that Aliria would fly to Boston again the next year for more scans and cognitive tests. Aliria's adult children were being followed as well. Someday, of course, the researchers could expect to have Aliria's autopsy brain, which might reveal further secrets of her resistance.

A BURST OF CHATTER on the brain bank's WhatsApp led me to realize that Piedad Rua's youngest sister, Camila, had died in Girardota. Piedad had reported the death right away to David Aguillón, Lopera's right-hand man, but then she'd left her phone in another room, lost in grief as she sat by Camila's body. Aguillón had called her fourteen times before they finally spoke.

The donation of Camila's brain was keenly awaited by Neurociencias. The first from the Girardota family, it promised to reveal more about how their mutation behaved. The WhatsApp messages from the brain bank staff and volunteers cited an extensive list of chemicals and supplies I'd never heard mentioned before. Camila's body was on the way to Medellín, where her brain would be subject to the elaborate preservation protocols that Kosik's pathologists had laid out earlier that year. Part of it would be flash frozen and flown in some sort of cryo-shipping box the next day—at enormous cost—to California, where Kosik's single-cell RNA sequencing studies would show which genes the mutation turned on or off in different cells.

It was a pivotal moment for the brain bank, which had continued to suffer major mishaps of late, including a collapsed shelf that spilled brains onto the floor. Camila's brain, preserved under the new protocols, would help provide the proof of concept needed before the NIH released more money for staff, equipment, and expertise. Aguillón and Andres Villegas barred me from attending the dissection of the brain, so anxious were they to avoid distractions. By dawn the next morning, the brain was on a plane to California.

Days later I saw Piedad in her home, where she, her family and her neighbors were on the fifth night of their traditional novena; nine days of evening prayers followed by a Mass for the dead. I got there early. In the garden behind her kitchen, Piedad was keeping chickens and a handsome pair of green parrots. I was used to her being harried, preoccupied, quick to complain. But she seemed at peace, exchanging baby talk with the parrots, tossing hunks of brown sugar into a large pot of water for the herbal tea she would serve later. Camila was now interred in the cemetery below the house, the sixth among Piedad's siblings.

We sat gazing out at the view from her porch, and Piedad pointed to a spot on a hill she identified as Paraíso, another of the historically Black hamlets of Girardota. She had a relative with Alzheimer's living there, she told me. "And another over there," she said, pointing to a different hill.

She asked me about the woman's case she'd been hearing so much about. When news of Aliria had reached Piedad, she'd called one of the nurses at Neurociencias to ask when a drug would become available. This was what people were taking from the story: a miracle case, an imminent drug. The nurse explained as gently as he could that, even assuming that the case did lead to a treatment, it would be years before any medicine resulted.

Piedad had been caring for Camila more than I realized—every day she did both a morning and evening shift—and she'd been the sole witness to her sister's death. Camila looked strange that day, Piedad told me. Her color was bad, she was vomiting and sweating, and Piedad carried her in her arms from one bed to another, trying to make her more comfortable. She had called Camila's children and husband to come quickly, but they did not make it in time.

Camila was fifty-five when she died. Neither her husband nor her children, who were in their twenties, were attending the novenas in Piedad's home. Young people responded in all different ways to their parents' illnesses and deaths, I realized; there was no script for this.

Late that afternoon, I joined the neighbors and friends who'd begun filing in and sitting down on the plastic chairs Piedad had arranged against the walls. Within minutes there were forty people in the house. Piedad's sister announced to those sitting around her that Camila had been her sixth sibling to die, and that all six were now in the cemetery. "I have three in the cemetery," the woman next to her responded blithely. That sparked a round of "How many dead do you have?" as others chimed in with their counts.

That six siblings should be dead before age sixty did not impress the people who had gathered here, and no one mentioned Alzheimer's disease. What was Alzheimer's but another way to die, in a place where early death was unremarkable? The novena began promptly at five o'clock, led by a volunteer from the church who recited, in a soothing baritone, the ageless Catholic prayers: the Glory Be, the Hail Mary, Our Father, and the Requiem.

At interims the group sang hymns, led by Piedad's strong, smoky voice. The sun went down and a cool breeze blew through the house. There was no food and no extravagance, just some people coming to say prayers for the recently departed. They would repeat the sendoff for four more evenings. Piedad ladled out cups of her sweetened herbal tea, and everyone slinked out the door. The first of the season's Christmas lights were sparkling in the hills.

TWENTY-TWO

Just weeks before her scheduled genetic test, Daniela López Pineda fractured her foot by stepping in a pothole. It had happened on her way to work. Daniela found herself housebound in Bogotá, on leave from her job, and surviving on the savings that was supposed to be used for her test, which she canceled. With the next available orthopedist appointment months away, she returned to Medellín, whose health system she found easier to navigate. She arrived on crutches, looking heavier than usual and in obvious pain. In her handbag was a copy of Gabriel García Márquez's *One Hundred Years of Solitude*, which had kept her company on the ten-hour bus ride.

I'd invited her to stay with me while she sought medical help, as I had more space than her sister. She set up in my spare bedroom, where the first thing she did after unpacking her suitcase was to tack up a photo of her mother, Doralba, in her trendy pre-Christian era. Within days Daniela had seen a specialist, who began treating her with steroid injections and sent her for physical therapy. Soon she was using a cane instead of crutches, but something was wrong with the way her foot had healed.

This injury was the last thing Daniela needed. Her new job did not pay well, but it had given her new friends—her colleagues had thrown her a twenty-seventh birthday party in her cubicle—and the sense of independence she craved. Her eight-year relationship with Ximena was now over, with her foot injury the breaking point. Ximena's indifference to it had made Daniela wonder what would happen if she one day developed Alzheimer's, and La Flaca seconded that thought. Daniela had only two months before her disability coverage ran out. She would either return to work with a bad foot, be fired, or manage to secure a surgery.

THE CHRISTMAS SEASON STARTED as it always did, with nightly fireworks from the first of December that kept dogs cowering and people awake in their beds. The staff at Neurociencias prepared for its usual year-end surge of brain donations, and Lopera had two parties scheduled, one in Yarumal and the other in Medellín.

Ken Kosik was back in Medellín, once again searching for the origins and meaning of the eleven Colombian presenilin mutations he had so far identified. In a few days Kosik, his student Juliana Acosta-Uribe, and Lopera would fly to the city of Pasto, near the border with Ecuador, to meet a family with a presenilin mutation who got dementia in their fifties. Families like this found Neurociencias now, not the other way around; their emails came from every corner of Colombia.

After three hours we pulled into Yarumal, where the families were waiting for their Christmas lunch. Lopera's agenda for the day also involved confronting a touchy issue: recently, representatives of a Colorado company called International Hemp Solutions had shown up in Yarumal to propose a study of marijuana in the E280A families, and they'd finagled a meeting with government officials to press their case.

The lunch was held in an overdecorated banquet hall; when we arrived at noon, dozens of people were waiting, most of them participants in the crenezumab trial who got their doses at the trial site in

Yarumal. Lopera launched in with "the good news and bad news" of 2019, firing up his slides of amyloid and tau. Yarumal was his hometown, and he was in his element: laughing, spirited, folksy.

As was his custom, Lopera delivered the bad news first. Trials of virtually every anti-amyloid therapy had been suspended. A couple were still being tested, he said, but things didn't look great.

But there were two pieces of good news, he said. The resurrection of Biogen's drug aducanumab, after early news of its cancellation, meant new hope for a drug that was similar to crenezumab. "We are more optimistic than we were in March," he said, adding that there would soon be results for two more antibodies, gantenerumab and solazenumab.

The second bit of good news—which Lopera did not downplay—was Aliria. Most of Lopera's audience had by now heard of the woman he now described to them as "born with the illness, and born with the cure." I recognized that line from the recent article in *El Tiempo*, and though it was quite a stretch to employ the word *cure* when talking about Alzheimer's, it was one he would repeat boldly in the years to come. As Lopera explained more about the Aliria case—her survival into her late seventies with only light impairment, her trips to Boston—the crowd in Yarumal grew wide-eyed. The finding, he said, had been published "in the most important medical journal in the world," and represented "the best news for us in thirty-five years."

Lopera moved on to the issue of genetic disclosure. Starting in 2022, he announced, all trial participants would be given the option to learn their mutation status. Lopera sounded firmer about this than he had in past years, and happier.

"I want to know right now," a man yelled out.

Lopera nodded. "In 2022," he repeated.

Finally, he addressed the marijuana study. People had come to Yarumal looking to test cannabis in the families, he explained. He didn't know much about the study, but anyone who enrolled would have to drop out of API Colombia, because "we won't know if it's the cannabis or the crenezumab" that is helping. He sounded entirely cheerful, but I knew that Lopera was privately furious about the cannabis project, which he viewed as poorly conceived and potentially undermining of his work. A lot of people in the families smoked marijuana; if cannabis prevented Alzheimer's in E280A carriers, this epidemic would have ended a long time ago.

THE NEXT DAY, in Medellín, Lopera gave a more detailed Christmas talk. The crowd at the SIU was larger than in Yarumal, and their questions were unusually pointed. The people I knew from the trial were present, but the ones asking questions were strangers to me. They'd been observing their fellow participants, and seeing that many had symptoms. A few were so badly off that they had trouble remaining still for their infusions.

"What percentage sure are you that this is working?" one attendee asked Lopera.

"We don't have any certainty, just hope," Lopera answered. "For thirty years we could only observe you and accompany you" in the disease process, he said, noting that the trial marked the first opportunity to change the course of the disease. But it may fail, he acknowledged.

Some of the knottier implications of the Christchurch discovery were not lost on this crowd, either. Why are we on an anti-amyloid drug, one woman demanded, when signs were pointing to tau as the bigger culprit? Was crenezumab effective against tau? Lopera responded that there was early evidence from other anti-amyloid drug trials to suggest that these drugs also reduced tau. This trial would reveal more about the matter, because investigators were measuring tau in the brain.

What happened if crenezumab failed? people asked. Would there be further studies of other, more promising drugs? Lopera told them they'd be the first to know.

These were the same people who came to every Christmas party, yet this year they seemed less passive. They sounded almost like the DIAN participants I'd heard in Los Angeles.

The party was the last I would see of Lopera, or any of the investigators or families, for a very long time. In a little-known city in central China, the COVID-19 virus was spreading and would soon confine us all to our far-flung corners of the planet. Future Christmas parties would be awkward affairs over computer screens, when they occurred at all.

DANIELA PROVED A WONDERFUL roommate, scrupulously neat and full of cheer, though she was always spraying something to cover up the smell of pot, and she would disappear in the night without explanation, like a cat. She made friends effortlessly, and my whole neighborhood came to recognize the smiling young woman who walked with a cane. When there was too much leftover food in the house, she wrapped it up in cellophane and delivered it to the homeless men camped on the edge of a nearby drainage canal.

She shuttled frequently to La Flaca's, and on Sundays she made the rounds of her ill family members in their homes. Like her sister, Daniela viewed these visits as sacrosanct, an obligation unique to their clan.

She found her aunt Mabilia doing better than when I'd last seen her. Mabilia had by now received a hospital bed from Neurociencias, and though she was still skeletal and sick, she was less agitated and sleeping better thanks to the medicines they had provided. Daniela also spent hours with her aunt Elcy in the nursing home, singing to her and feeding her spoonfuls of gelatin, one of the few things she could eat anymore. Elcy was developing a bend in her neck that impeded the passage of food. After years in bed, Alzheimer's patients often had musculoskeletal deformities, and Daniela sensed that Elcy's end was near.

I accompanied her to visit her uncle Fredy, who was fifty and lived with his wife, daughter, and son in a threadbare apartment in Pedregal, the same neighborhood as his siblings Mabilia and Jaime. Fredy had the giant López forehead, crude tattoos on his forearms, and pale green eyes. He had been ill a full decade, but was not yet bedridden. He still walked and used the bathroom, with help, though no one would let him anywhere near the front door. Fredy spent most of his day seated on a sofa whose synthetic leather he peeled off in chunks, handing them to his family members like prizes.

Daniela had a sort of genius in getting through to Alzheimer's patients. To provoke a response from her uncle, she started reciting familiar names, loudly. For a while, as we sat with him and chatted, Fredy seemed just a mildly disoriented person at a party he was enjoying. He had only one word left—*Sí!*—and fragments of others that were unintelligible.

Fredy's wife had thick bobbed hair, a country accent, and the same harried energy as Piedad Rua and other long-term caretakers I'd met. She entered and exited the kitchen as we talked, telling us—first with a plantain in her hand, then a carrot, then a giant knife—that she didn't want Fredy to die. For years they had lived like this, with Fredy dependent and unable to work, their income the sum of his meager pension from a construction firm and whatever the kids brought in. Yet she loved him as much as ever, she affirmed.

Like his sister Doralba, Fredy had been born with a wild streak. His wife believed that Fredy's partying had triggered the early onset of his illness, at age forty. Their twenty-seven-year-old son, Bryan, shared his father's tendency to drink and use hard drugs, causing his mother angst. But Bryan was good with his father, feeding him cake and yogurt as we talked and escorting him patiently to the bathroom, holding him by both hands for balance. I caught glimpses of Bryan's glamorous sister in her bedroom, brushing her long shiny hair as she got ready to go ride around all night on the back of a motorcycle. Daniela had explained to me that she was a gangster's girlfriend, which was sort of like having a job, and that the boyfriend's money helped support the rest of them.

Why doesn't Lina ever visit us? Fredy's wife asked Daniela, who struggled to answer. It was hard to explain why her cousin refused to see her sick aunts and uncles. And what was happening with María Elena? When she'd visited recently, Fredy's wife said, she was repeating herself a lot. No one wanted to think that María Elena, who was so supportive—she'd not only gotten Elcy into a good nursing home but had helped Fredy secure his pension years before—could become sick. Daniela did not volunteer an opinion. But I knew she'd just been to María Elena's house, and she was disturbed when her aunt asked her three times to grab her a tomato, not realizing that she'd just taken one and chopped it up.

EARLY JANUARY WAS RELENTLESSLY hot in Medellín. After the New Year, a sort of collective hangover descends on the city, and it was in those still, lethargic days that Daniela, faced with the threat of losing her job, returned to Bogotá. Physical therapy and injections had improved her foot, but she still limped and suffered, and finally she was put on a list for surgery. She didn't know if it would be a month or a year before she could come back to Medellín for it. She rented a room and did not tell Ximena she was back.

I visited La Flaca and her husband, Fernay, who, tired of driving a cab and frustrated at his inability to save more money, had recently opened a small gold mine with a group of friends. He explained how a tiny bit of mercury goes into a drum full of broken-up ore mixed with water, and all the gold sticks to it until it's a glob. He retrieved from his pocket a heavy round pellet with a sulfur color and matte texture. This was unrefined gold, he said, straight from the drum.

Flaca had a headache and was resting her face in her hands, waiting for the ibuprofen to kick in. An enormous new painting of Jesus dominated their living room.

Their life was entering a joyful phase. Their son was doing well at soccer, was top in his class, and had just made his first communion. Fernay and La Flaca had never married in the church, for lack of money, but now they planned to do so; Fernay showed me the gold bands they'd picked for the ceremony.

I wondered whether Fernay's aggressive push to earn money wasn't on some deep unspoken level about the specter of Alzheimer's—a bid to provide for his family in the event that luck went against them. A small, private gold mine in rural Antioquia was a far riskier venture than he was making it out to be. Extortion by armed groups, accidents, and all kinds of other bad things happened. But his excitement over that little sulfur ball was infectious.

ON SATURDAY, JANUARY 11, Elcy López Pineda died. Usually, the brain bank WhatsApp chain was how I found out about a death, but it was La Flaca who alerted me this time.

The family had long ago agreed to donate Elcy's brain to Neurociencias, but the nursing home had forgotten that instruction and sent her body to a public morgue. Elcy had been dead twelve hours by the time her body was moved to the funeral home near the SIU, where David Aguillón and his team arrived with their white bucket in the intense noon heat.

Elcy's withered form was curled on the steel table when I stepped into the room. She looked as though she had starved; her skin hung from the bones of her arms. She had survived more than a decade with the disease and never suffered seizures, Aguillón knew from her records. That meant her brain would be an interesting one to study. But it had been sitting too long, on a hot day, to produce good results for single-cell RNA sequencing or electron microscopy. After Muñeton, the prosector, arrived in his scrubs to open Elcy's cranium, Aguillón decided that Elcy's brain would go whole into preservative, none of it sliced and frozen. The brain weighed 776 grams, just more than a pound and a half. Like everything about Elcy, it was tiny and wasted.

La Flaca, who had changed the photo on her WhatsApp to a black ribbon of mourning, wrote to ask how Elcy had looked. "Very, very skinny," I said. Elcy's body was to be cremated, and the funeral Mass held the next day. Who would come? I wondered. Elsy had been gone from the world so long that a funeral was almost redundant. At least she had seen one familiar face—Daniela's—before she died.

THE ONLY PEOPLE TO attend Elcy's funeral, in the basement of a church in Pedregal, were family. Cousin Lina showed up with her sister. La Flaca and Fernay arrived by motorcycle, the two of them smiling and joking. María Elena came with a daughter, and their uncle Jaime showed up after the Mass had already started.

Two funeral home attendants in gray uniforms placed Elcy's ashes on a stand before the altar. Street noises—music, hawkers, and some boys descending the hill on horses—invaded the church basement as the priest delivered his generic homily, in which exactly nothing was said about Elcy López Pineda. The whole thing was over in half an hour. Everyone remained dry-eyed until the funeral girls handed María Elena the white box containing Elcy's cremated remains, and María Elena opened the lid as though compelled to suffer their full, terrible impact. She promptly broke down weeping, and had to be held.

Afterward, the group milled around the church for a few minutes before dispersing. No one offered or suggested lunch. There was no bringing the family together to talk about Elcy, or to talk at all.

When I called Daniela to report on this curt, chilly funeral, none of it surprised her. The López Pinedas had been through too much to enjoy one another's company anymore, she told me. They were so divided by class and money, by disease and fear and resentment, that they could no longer stand to be together as a group. Each death caused them to relive the trauma of a previous death, and raised the specter of future ones. The older members of her generation—Lina and Flaca—were already in their thirties, Daniela reminded me; this was not so far away.

Exactly one week later, her aunt Mabilia died.

TWENTY-THREE

In 1907, Alois Alzheimer had pointed to neurofibrillary tangles as one of two signature brain lesions in his "peculiar disease." But eighty years passed before Ken Kosik and his colleagues identified the protein in those tangles as tau, and another three decades before the Alzheimer's field came to terms with tau's importance. Until about 2015, most researchers had remained laser-focused on amyloid. But the failures of several anti-amyloid drugs, combined with a flurry of discoveries about tau, caused some scientists—including longtime champions of the amyloid hypothesis—to start taking tau more seriously.

What happened in Alzheimer's, and several other neurodegenerative diseases, was that tau, which is a normal component of neurons, underwent chemical changes that caused it to misfold into an abnormal structure, like a smooth sheet of paper crumpling. This misfolded tau was able to move out of the cell and into a neighboring one, causing its tau to misfold in the same way. The bad tau became self-propagating, eventually spreading through the brain.

In Alzheimer's disease, tau was long deemed to have a supporting role to amyloid. But the relationship between tau tangles and symptoms was clearer, with higher levels corresponding to worsening function and brain atrophy in Alzheimer's patients. Some early work in autopsy brains had illuminated these links between tau and symptoms, but tau now could be measured in living people as well, through new blood tests and nuclear imaging. Yakeel Quiroz's studies in E280A carriers had shown that small, incremental increases in tau tangles were linked to poorer performance on memory tests among people still in their thirties.

In 2017, David Holtzman of Washington University in St. Louis had stunned the Alzheimer's world by showing that the risk gene APOE4, already known for stimulating amyloid in the brain, had a profound impact on tau that was independent of its effects on amyloid. Holtzman and his colleagues found that interactions between APOE and tau sparked a dangerous immune response in the brain that led to the death of neurons. But an array of key questions continued to elude researchers: What, exactly, caused tau to misfold? Was it tau or inflammation causing neurons to die? Was there synergy between amyloid and tau? While most researchers still considered amyloid relevant to Alzheimer's, by 2020 they were convinced that cracking tau was also going to be fundamental to treating the disease.

Kosik and his colleague Jennifer Rauch had recently discovered that a protein in the cell membrane, called LRP1, acted as a "master regulator" of tau uptake and spread. If you suppress LRP1, Rauch and Kosik found, tau cannot move effectively from neuron to neuron. The Christchurch finding, whose significance Kosik had initially been skeptical about, now seemed more plausible, because APOE was also known to interact with LRP1. Was it through LRP1 that Christchurch had hampered the spread of tau in Aliria's brain?

Researchers already had a number of drugs in development that targeted tau. There were antibodies designed to clear tau, gene-silencing therapies, and drugs to stop tau from aggregating in the brain. The field wanted to move ahead with tau treatments, but didn't want to repeat the expensive, demoralizing mistakes of the anti-amyloid drugs, with doses that were too low or too high, antibodies that targeted the wrong form, and uncertainty over which stage of illness was ideal to begin treatment.

ON VALENTINE'S DAY, I visited La Flaca at her home. It was a weekend when Fernay was away at his jungle mine. He sent photos of boa constrictors that slithered their way onto the site, and the primitive tunnel mouth into which the miners descended night and day.

I hadn't seen Flaca since Mabilia's death a month earlier, and she wanted to know what I'd seen.

Mabilia had died at eight o'clock in the morning on the Saturday after her sister Elcy, who'd been sick twice as long. No one had expected Mabilia to go so soon, but she'd been extremely ill the week before, likely with an infection. Her family had already agreed to donate her brain when she died. When I got to the SIU to wait for David Aguillón, I saw that Elcy and Mabilia had been assigned consecutive numbers at the brain bank: no one had died in the interim.

Mabilia's body, unlike Elcy's, had arrived in good time, meriting all the high-tech protocols the brain bank now followed. The procedure was orchestrated by Aguillón, who, though he was not yet thirty years old, had absorbed, Jedi-like, decades worth of his older colleagues' institutional knowledge. By that time, he had presided over one hundred and eighty donations. When I arrived, his students were printing tickets to stick on the baggies and vials, and the prosector was on his way.

The tensest part of any autopsy was the paperwork. A handful of López Pinedas, including María Elena, Lina, and Mabilia's son, Sebastián, were seated waiting around one of the funeral home's glass tables when Aguillón and I walked in. It was Sebastián's job to sign the papers, and he was brave and polite as he began to review them.

Only when Aguillón ventured to reassure Sebastián that his mother's death was due to the disease, and not any failure of his care, did Sebastián's reserve finally crack, and he cried. Lina suddenly looked flushed and furious. Though Aguillón needed his papers signed, Lina held him up with a flurry of angry questions: Why did the people Neurociencias called looking to study always end up getting sick? she demanded. Would they be able to find out if anyone in their own family had the Christchurch mutation? When would they get to learn whether they carried E280A? And what about the money for the brain?

Aguillón kept his cool with Lina. The money they could pick up during the week, he replied. It would be three to five years before they would start giving out genetic results, he told her, and so far only one person had been found to have two copies of Christchurch. The researchers did not know who was an E280A carrier and who was not, he told Lina.

María Elena looked mortified at this fractious conversation; this was not how her generation dealt with Neurociencias. Sebastián signed the papers. I said goodbye to the aggrieved López Pinedas and followed Aguillón's team back through the steel swinging doors of the mortuary, where Mabilia's body was waiting.

In the dissection lab an hour later, Aguillón had his students guess at the weight of Mabilia's brain just by sight. It weighed 915 grams. Mabilia died nearly as skinny as Elcy, but she had more of her brain left, as she'd been ill only half as long. She had not yet lost her speech at the time of her death, and beckoned her son by a name she could no longer pronounce in its entirety: *Seb. Seb.*

I later asked Lina what had bothered her at the funeral home. She told me that Sebastián had been trying to get Aguillón on the phone for days before her death, as his mother grew feverish and her condition worsened. But Aguillón never returned the call. Sebastián was savvier than either of his parents, Lina said—Mabilia had only a second-grade education—but he'd never learned to work the health system. Like many E280A families, they could access a doctor only with difficulty, so they called Neurociencias instead.

Sebastián was still upset with Aguillón when the family had arrived at the funeral home to sign the papers releasing Mabilia's brain. But when Aguillón started reassuring Sebastián that Mabilia's death wasn't his fault, Lina recoiled at what she saw as manipulation. The researchers, in Lina's eyes, only cared about brains, blood, and data. She hated how her aunts fawned over them: *yes doctor thank you so much doctor.* As a pharmacist, Lina said, she knew that if crenezumab was found to work, the families would never get it. "They'll just be used as guinea pigs for five years and after that, 'Thanks. Bye,'" she said.

HOW HAD MARÍA ELENA acted at the funeral home? Flaca wanted to know.

I thought about it. María Elena hadn't interacted with me, I realized, either there or at Elcy's funeral the week before. She hadn't spoken to me or even looked at me, for that matter, though we'd had long, friendly conversations in the past.

I chalked this up to grief, but Flaca said that wasn't it. "She thinks you know whether she's getting sick. She thinks you have information from Neurociencias."

I did not, I assured her. All I knew was what her nieces told me.

Flaca said she was worried about her aunt. The family had decided to use the money from the two brain donations—Elcy's and Mabilia's—as a down payment on an ossuary to hold the remains of all the López Pineda siblings who had died. They would put Doralba, Amparo, Nicolás, Elcy, and Mabilia together, and there would be room for more to come. María Elena and Flaca were in charge of the effort. They arranged to go into Neurociencias together on a Thursday and pick up the cash. But then María Elena had called Flaca on Wednesday to ask whether she had gone to get the money. That struck Flaca as strange. And María Elena had stopped going downtown to shop, even though she'd always made trips once or twice a week. She must have gotten lost, Flaca thought.

María Elena was acting just like Doralba in the early stages of her disease: confused and slightly paranoid. It didn't help that María Elena's family was not supportive. They had browbeaten her about participating in the Neurociencias studies so badly that she'd eventually dropped out.

TWENTY-FOUR

The coronavirus pandemic reached Colombia in March 2020. For a month the country had waited tensely for the virus to hit, knowing it was a matter of time and that a strict quarantine was inevitable. Medellín's itinerant fruit vendors capitalized on the uncertainty by boasting of the antiviral properties of their oranges, and its herbalists claimed that a plant called moringa, prepared as a tea with limes and raw cane sugar, could steel the immune system for the impending onslaught.

I spent the days before the lockdowns with Blanca Nidia Pulgarín and her family. The first Alzheimer's clan Lopera ever studied, they were the only family I knew who had maintained their traditional ways all these years. Blanca Nidia urged me to visit amid warm, dry weather and a security situation she deemed tolerable.

From my bus to Belmira I saw that buildings along the highway were splattered with the graffiti of the fearsome gang of drug lords and ex-paramilitaries who controlled vast territories from Medellín to the coast. The crater-faced driver who ferried me in a 4x4 from the bus terminal to Blanca Nidia's home made no secret of the fact that he worked with them. "Now I have you kidnapped," he announced in the middle of that slow, bumpy journey, placing his hand on my thigh. I tried to treat it as a joke, and I arrived unscathed. The next year

Blanca Nidia would report to me that the driver—with whom she'd also had to travel on return trips from Neurociencias, when she couldn't catch a ride with a milk truck—was dead, shot in the back multiple times.

Blanca Nidia lived on the farm she'd inherited from her father, Pedro Julio. It was a working finca of simple design, with thick earthen walls and deep-set windows with wooden shutters. Like the other farmhouses in the hamlet, hers sat on a little cleared platform in the sky, against a backdrop of wild dark woods, and its view extended all the way to the sinuous Cauca River, which caught a burning glow as the day progressed. In the afternoons, when all traces of fog had dispersed and a gold light blanketed the hills, the view from Blanca Nidia's porch became a moving painting; I squinted watching a distant horse inch its way up the terraced, almost vertical slopes. The family did not own a pair of binoculars but could distinguish with their sharp eyes whose animals and milk cans were on the march.

"Listen to this," Blanca Nidia said. The radio was playing news of a cocaine lab destroyed days earlier in a lower hamlet, through which I'd passed to get here. The lab was producing two tons of cocaine a month. Nine people had been arrested.

Blanca Nidia and her husband listened to the radio all day. Cell phones couldn't catch a signal at this remove, except at one spot just off the kitchen that was marked with a hanging string. There was no internet. They kept one radio going at high volume in the kitchen and brought a portable one with them when they went to milk the cows, listening to rural variety programs that combined news, music, and public service announcements. As a result they were better informed on politics, crime, and the looming pandemic than their relatives in Medellín, whose news was heavy on rumor and WhatsApp innuendo.

Their twenty-three-year-old son was committed to living alongside his parents, on this wedge of mountain, and making his life here. He left the hamlet only to buy groceries every two weeks, or to see his partner and their daughter. His safety depended on his staying put as much as possible, Blanca Nidia explained. A cross still marked the spot where his older brother was found dead after having stumbled across a lab.

The Pulgaríns kept twenty cows, a dozen or so rabbits, two goats, scores of chickens and turkeys and ducks, several calves, a mare, and a mule. Twenty cows is full-time work for two men. It took two hours every morning and evening to call each cow to the milking tent by her name—*Karen! Martica! Yolanda!*—feed them, milk them, and disinfect their teats. In the middle of the day the men repaired electric fences, cleaned milking equipment, cut tumors off udders, and performed countless other tasks that left them only an hour after lunch to rest.

Blanca Nidia's day was no less regimented. She awoke at four, showered in cold water, slapped on some makeup and her big silver cross, and tied up the lace curtains that separated the bedrooms of the railcar-like house. She brought coffee to the men as they milked, started breakfast, swept and mopped and watered her plants, and chopped up potatoes for lunch using her bare hand as a cutting board. In the afternoons she would grind cheese for the evening's *buñuelos*, collect firewood, and prepare a huge cauldron of boiling water to clean the milk cans with. The family got everything done before dark, then retired by eight or nine o'clock.

They yelled to one another nonstop. It was the only way in these hills. Blanca's high clear voice reached so far that other people sometimes asked her to yell things for them.

I figured that Blanca Nidia, now sixty-three, was out of danger when it came to E280A, that she was a noncarrier. But I knew what Alzheimer's disease had done over the years to farming families like hers. Sonia Moreno had told me once that farmers were often diagnosed late, because their lives were so repetitive that they could maintain their daily routines until well into its course. I had a hard time believing, after seeing the hyperdisciplined way this family lived, that any type of disability or incompetence could go unnoticed very long; even Blanca Nidia's husband only ever let himself get drunk once a year, at Christmas. People weren't having nearly as many kids as they used to, and this meant that the farms were chronically shorthanded. The labor of one man or woman was worth a lot, and not easily replaced.

At the moment, no one in the hamlet had Alzheimer's. Blanca Nidia's sick siblings were being cared for in neighboring towns. But E280A surely lurked. The children of Blanca Nidia's sister, brother, and cousins who had fallen ill or died of it in recent years were not yet of age; the disease was on pause. Somewhere on this slope, stashed on one of its farms, was an unused hospital bed from Neurociencias. There was no hurry to return it, as sooner or later someone was bound to get sick.

On the day before I was to leave—Blanca Nidia's grandson agreed to fetch me on his motorcycle so I could avoid that sinister driver—Blanca Nidia heard the distant rumble of a car headed toward her finca. I couldn't hear it, but she could, and soon enough a boxy old Toyota Land Cruiser rolled up. A man hopped out to warn that there were people on the way. One of Blanca Nidia's aunts, who lived across the valley, had fallen ill and was being carried here on a *palo*, which could mean a few things: a pole, a trunk, or a stick. The car would wait here at Blanca Nidia's—one of the few farms with road access—to take her to the hospital.

Blanca Nidia, unfazed, announced she was making a pitcher of lemonade because she was sure we could expect "lots of people." Half an hour later, a dozen men arrived, on horses and on foot. Two of them were carrying a body that was wrapped in blankets and tied at each end to a narrow tree trunk some four inches across. They had walked for hours, through woods and streams, up and down hills, with the weight of this human suspended between them. Carefully they lowered the trunk to the ground and untied the blankets, revealing a heavy gray-haired woman who struggled to sit up. When they lifted her to help her toward the bathroom, she appeared completely limp on one side. Blanca Nidia said she suspected a stroke. Her aunt's health had been precarious for a long time, she said, but she had refused to see a doctor.

Within minutes two of the men had loaded the aunt into the Toyota and set off for the nearest hospital, two hours away. The rest of the posse lingered a while longer, smoking and chatting, letting their horses graze. Blanca Nidia's son seemed glad to see them; most of these strong young farmers in rubber boots were his cousins. The men of this hamlet had been volunteering themselves to carry the sick, the dead, and people in the end stages of dementia on *palos* as long as anyone could remember. "They're very humanitarian," Blanca Nidia said proudly.

ON MARCH 25, quarantine went into effect across Colombia, and silence descended over Medellín. That evening a helicopter began circling low and loud above the city, broadcasting a man's voice through a crackling loudspeaker. The words were distorted, but had something to do with coronavirus: a benign, insipid message delivered in alarming fashion to a city full of refugees from conflicts in the countryside, people with bad memories of curfews and helicopters. The government had opted on a militant approach to the pandemic. Otherwise how would Medellín's population, nearly half of whom worked off the books and in the streets in what was euphemistically called the "informal" economy, be persuaded to stay home? There was no talk yet of government aid, and their houses—often simple, utilitarian, homemade constructions of porous brick—were not built for staying indoors all day, much less in the company of one's extended family. Every night for weeks, the helicopter circled and crowed, though the voice changed from a man's to a woman's.

In early April, Medellín still registered fewer than two hundred cases of COVID, but no one expected these low numbers to hold for long, and hospitals scrambled to add beds in anticipation of more. A month into quarantine, the city was still depressed, trapped in a listless calm like one of the strange dips of energy it always suffered after Christmas.

THE COVID LOCKDOWNS threw a wrench into clinical trials worldwide, and API Colombia was no exception. The crenezumab infusions were halted, and no one knew when or whether dosing would resume. It had been less than a year since patients had been switched to high-dose intravenous infusions, and investigators worried that the pause in dosing could make the eventual results even harder to interpret. Francisco Lopera, who would turn seventy the next year, was remanded to his finca by the university for his safety. Lopera made all the important decisions for his group, yet there was no definite plan for his succession. He and his colleagues all knew that if he were to contract the virus and die, Neurociencias's work—and that of its collaborators abroad—would be in serious trouble.

The investigators had long sent gifts of groceries to households they considered to be in urgent need, but this described most of them now. The social program's yearly budget got diverted to food and medicine aid. David Aguillón took charge of the deliveries, because physicians had blanket permission to move around the city. The aid was a drop in the bucket. But the Colombian government was hardly doing better. The specter of hunger grew darker. In my middle-class neighborhood, people started ringing random doorbells in search of food, many of them young women hoping for baby formula. Suddenly there were reports of people forming roadblocks and plundering grocery trucks.

While the shopping malls and residential sectors remained silent, the city's downtown emerged as a sort of tolerance zone, an escape valve for hundreds of thousands of people who needed to buy and sell goods in real time, or they would starve. Cloth-mask vendors, their products stitched by an army of laid-off seamstresses in their homes, made the rounds with their pushcarts. The adaptability of the paisas, and their relentless commercial instincts, had saved them from hunger, if not the virus itself.

In late May, API Colombia resumed dosing. Participants received special permission to travel to the trial center, where they faced exacting measures of masking and distancing. Still, many caught the virus, and the center was forced to close repeatedly. Lopera, still confined to his farm, delegated his major duties to Aguillón while he spent his days feeding his dogs and birds, and attending Zoom meetings, and letting his white hair grow long. COVID was just the latest existential threat to a trial that had already seen plenty.

DANIELA CELEBRATED HER twenty-eighth birthday alone in her rental room in Bogotá, where she'd been calling her sister several times a day since quarantine began. While her life had changed in the two years since her mother's death, it had not much improved. She'd extricated herself from her unhappy relationship only to suffer with an injured foot. She had now reached the age when plaques can first appear in the brains of E280A carriers. If she did have the mutation, and planned to live as full a life as she could before it struck, she didn't really have a day to waste.

Flaca urged her sister to return permanently to Medellín, in part to keep her company. Flaca's husband remained camped out at his rural gold mine, where he had recently been thrown into jail by police demanding a cut of the proceeds. Teenagers with guns and ties to mafias were also showing up to harass the miners. As dangerous as Fernay's situation had become, and as much as he risked seeing no earnings whatsoever if this continued, there was no point in driving a taxi with streets still largely empty.

In June, Daniela was finally let go from her job, which eased her decision to leave Bogotá forever. She double-masked herself and boarded a bus to Medellín, with a cane in hand and all her possessions stuffed into two flimsy suitcases. Intent on getting her foot surgery while she still had insurance, she set out the next day to tackle the health bureaucracy. COVID cases ebbed and flowed in Medellín, allowing hospitals to offer procedures when levels were low. Soon thereafter, in a rare break, Daniela received notice that her operation had been approved. She went in for surgery just before the hospitals shut down again. Flaca cared for Daniela in her home, cooking for her and helping her bathe.

COVID was everywhere. As quarantine relaxed—not by government decree but attrition and desperation—cases surged. Even Blanca Nidia managed to get the virus in her remote mountain hamlet. Piedad Rua of Girardota, a heavy smoker in her late sixties, barely survived her bout. Andrés Villegas, the brain bank's director, spent weeks hospitalized in intensive care. Soon the geneticist Gabriel Bedoya, too, would be hospitalized, never to see the inside of his lab again.

At the SIU, which was still locked down and off limits to outsiders, brain donations continued with difficulty, under a protocol that involved coveralls with hoods, goggles, and the N95 face masks that were so scarce in those days. But the brains were more abundant than ever. Without access to symptomatic treatments, and with their families' former breadwinners less able to provide for them, the Alzheimer's patients were declining faster.

IN LA CEJA, Marta Pulgarín Balbín struggled to keep her mother and aunt safe from COVID. In the United States, Marta's sister Paula found herself on the front lines of COVID, working long hours with the sickest patients. She kept in close touch with her siblings throughout the pandemic, and sent them money. One night she cried on the phone with Marcela, for whom the scarcity and uncertainty of the lockdowns brought back memories of their hungry childhood.

Before COVID, Paula had been moving forward with a plan to test for the E280A mutation. She had consulted a neurologist to help her navigate the process.

Paula, who had one biological child, had hoped to adopt a second child, just as her parents had done with her. She and her husband had already started the paperwork. But if Paula's genetic test were positive, they decided, they would halt the adoption. Paula took out a costly long-term care insurance policy, knowing she would not be able to do so after a bad result.

IN THE UNITED STATES, the Alzheimer's Association held its annual conference online, denying both Lopera and Yakeel Quiroz the full excitement of the awards bestowed on them that year. Lopera was recognized by the association for his life's work, and Quiroz for her important work on the Aliria case and Christchurch mutation.

Marta Pulgarín Balbín left me long messages in her honeyed voice to update me on the suspensions of the clinical trial, which had everyone on edge. One day in early September, when I happened to call her to check in, she told me that her Aunt Elida had died just hours earlier. With COVID cases not too high in La Ceja, a funeral would proceed the next day, she said.

I arrived the next day to find the family sitting around the casket in a ring of chairs, their surgical masks dangling below their chins, saying a final rosary before the funeral Mass. Little kids played on the floor, and a sign forbade people from drinking alcohol or playing music during wakes.

Mery's gaze was less focused than the last time I'd seen her, but she was still mobile and agile. As Marta escorted her mother to peer at her dead sister's face one last time, Mery did just as she'd done at the airport with Paula. She collapsed and had to be carried and fussed over. Some sixty people made it into the church soon afterward. Most of Antioquia's Catholic churches had cautiously reopened, with a lot of sanitizer sprayed and strict distancing between family groups. To have sixty people gathered during a pandemic was a remarkable showing of solidarity by this family that had once been so troubled, and quite a contrast to the López Pinedas, who had barely spoken to one another at Elcy's funeral.

The priest said nothing about Alzheimer's, or Elida for that matter, because they never did. He talked about forgiveness, in keeping with the Gospel readings for the day: "Then Peter approaching asked him, 'Lord, if my brother sins against me, how often must I forgive him? As many as seven times?' Jesus answered, 'I say to you, not seven times but seventy-seven times.'"

Marcela rubbed her mother's back tenderly, and Marta wiped her tears.

When the Mass had ended and Elida's body was loaded into a hearse bound for the crematorium, the mourners threw COVID caution to the wind and reconvened in Mery's house, which was still a mess from the chaos of the previous day. Marta explained to me that Elida had aspirated as Marta tried to feed her. Marta put in several panicked calls to Neurociencias, to no avail. Elida ended up dying in Marta's arms. Even as Marta went ahead with funeral arrangements, no one from Neurociencias called back. There was no talk of a brain donation; Marta felt let down by the researchers and was not about to bother.

As a relative sent out grilled bologna and cheese sandwiches from the kitchen, Mery, suddenly chipper, sat up straight on her departed sister's hospital bed and launched into a story. It was about a job she'd had in a La Ceja cantina, where she'd walked in for her interview to find "chicks with their tits out everywhere." Mery seemed to have forgotten the reason we were gathered. Everyone forced a half chuckle at her stories except her son Carlos, who stared enigmatically at his mother, as though contemplating a deeper horror. "It's five now," he said to me during a pause in Mery's monologue. It took me a moment to understand what he meant: five of his mother's generation were now dead from Alzheimer's. Mery was poised to become number six.

DANIELA, AFTER A MONTH with Flaca, craved her freedom, and had come to re-occupy my spare bedroom. It was nice having her around to recount to her the stories of other Alzheimer's families, and see how these resonated with her experience. I told her how Paula had purchased long-term care insurance, and Daniela said she'd never heard of it. For people like Marta's aunt Elida, or her mother, insurance was having a person in whose arms you might be lucky enough to die.

I told Daniela that Neurociencias had not responded to Marta's distress calls, though Marta was a faithful participant in the crenezumab trial and had a good relationship with the investigators. It was the luck of the draw what help people received when the researchers were doing double duty.

The group tried to care for the families through its social program, which was dear to Lopera's heart. It was not well funded or staffed; Lopera hired its first and only social worker just the year before. But the *plan social* was something Neurociencias had developed long before there was outside money, and not to fulfill some legal or ethical mandate but because they believed in it. As API Colombia got off the ground, Lopera pressured the Banner scientists to foot the bill for the painting classes, youth workshops, hospital beds, medical ID bracelets, groceries, and caregiver training the group offered the E280A family members.

The social program's budget represented less than one percent of what was being spent on the API trial, yet Banner had agreed to fund it only after mulling whether its offerings could be considered coercive. This was how the clinical trials world had come to think: that the barest forms of assistance could be construed as enticements. It wasn't without reason. The previous century had seen egregious examples of coercing and misleading study subjects, especially poor ones. But the same drug firms also never hesitated to tempt doctors, paying them liberally as consultants or speakers. The Banner scientists, like many top Alzheimer's researchers, each saw considerable earnings from consulting fees.

Daniela felt that the concerns about enticement were overblown. What the Alzheimer's families needed, she argued, was a formal center of attention, a place with a dedicated staff to whom they could turn at any hour, where their calls would always be answered. The families required medical advice, nursing care, consistent help navigating the health and legal systems and getting the sick into homes, programs to get people employed again or back in school after they'd lost years to taking care of a parent. "We're a bunch of montañeros," Daniela said. "We don't know how to do this stuff."

She could use some help right now, she conceded. She had quit college when her mom became ill, mopped floors for a year, landed a professional job that paid only slightly better, and lost it. Now she had no idea when companies would start hiring again or if she would even be eligible. Her foot was still healing, giving her little chance of passing the fitness test all the companies insisted on. Daniela wore plus-size clothing, and that also hurt her chances in an image-obsessed society with a large pool of young people looking for work. If you couldn't fit into the company uniform, she knew from experience, you might as well not apply.

TWENTY-FIVE

It was October 2020, nearly a year since the paper describing Aliria's remarkable resistance to the E280A mutation, when I learned that she was gravely ill and hospitalized. All my efforts to see Aliria thus far had been thwarted by COVID, which was far from abating in Medellín. I wondered whether she might have become yet another victim of the virus.

When I ran into David Aguillón at the SIU, he revealed that Aliria was dying of melanoma, not COVID. Her cancer had metastasized during the lockdown, when health services were hard to attain. Aliria's late-stage diagnosis came as a shock to the researchers, who had expected to be studying her for years to come.

No patent application for a drug had appeared so far. But Yakeel Quiroz's husband, Joseph Arboleda-Velasquez, had solicited and received a grant from a Silicon Valley foundation, Open Philanthropy, to further his investigations into Aliria and other E280A outliers. This rare gift, which at $4 million was seven times larger than the average NIH grant that year, had been awarded on the strength of the Aliria case. Though the important experiments had yet to read out, Aliria's impact on the Alzheimer's field was already significant. The case had galvanized interest in a different type of approach, away from the relentless targeting of amyloid-beta and toward a deeper exploration of APOE and different cellular mechanisms in Alzheimer's disease.

ALIRIA'S CANCER WAS DEEMED untreatable. She was released to palliative care in her home. Her government-subsidized insurance didn't offer much, so Aguillón was visiting frequently. He had also brought in a pain specialist. This level of attention was unusual for an E280A patient, as was the prompt arrival of a hospital bed and the subsidy that Aliria's daughters had been receiving from the university's charitable foundation to help them care for their mother.

I reiterated my desire for a visit—it was now or never, I argued—but Aguillón explained that he couldn't just drag me along. It was imperative that he obtain Aliria's brain when she died, a brain that the researchers hoped would tell them more about how she had resisted E280A for so long. It was not an ideal time, he said, to introduce new people to the family. Everything about Aliria required an extra level of diplomacy, it seemed. I called Quiroz and Lopera, pleading to meet the woman who had lit up the Alzheimer's world, on whom so many hopes had been pinned. They agreed, and Aguillón, along with two more members of his team, took me to see Aliria the next day.

Barrio Pablo Escobar, straight up a hill from downtown Medellín, was marked at its entrance by a startling mural of the dead capo, his menacing enlarged face against a backdrop of the city at dusk. Except for the mural, the barrio now looked like any other working-class neighborhood in Medellín. Aliria's daughter Magaly lived up a steep, narrow alley of stairs topped by a thick web of power and cable lines.

Magaly's house was well appointed, with its gated porch facing the alley, shiny floors, paintings of fruit, and statues of saints. Her adult sons and grandchildren wandered in and out. A no-nonsense woman in her fifties, Magaly poured us cups of juice, then dispatched us straight back to the spare white room where her mother was lying on her hospital bed.

Aliria weighed perhaps eighty pounds. She was very fair, with her shoulder-length hair dyed auburn along with her eyebrows. Her family encouraged me to pull up a chair next to the bed rail and take off my surgical mask; at this point they were no longer worried about Aliria getting COVID. Aliria smiled and grasped my hand. This pixie-like woman in full makeup and sparkly jewelry, with a rosary around her neck, was "the case" Eric Reiman had been crowing about to scientists in Los Angeles the year before.

"I'm full of little balls," she confided, sounding as though she were passing on a juicy piece of gossip. She was talking about her melanomas, the dark eruptions on her chest and neck. It was cancer that had Aliria bedridden; her Alzheimer's symptoms were still quite mild. Before the pandemic she had maintained her own home, down the street, and still cooked her own meals. In December she would be seventy-eight, if she lived to see that birthday.

She was bright-eyed and warm, not letting go of my hand. I learned only the barest details of her life that day, as her memory was now impaired enough that point-blank questions tended to stump her. Instead Magaly, the clinicians, and I chatted casually by her bed, allowing Aliria to follow along and interject when she saw fit. *Eavemaria!* she would chime in cheerfully, or *Oiga!* Listen! Among the bits of information I gleaned was that she had grown up on a farm near Canoas, the hamlet of Angostura where E280A was so concentrated. She'd had almost no education, and she'd left for Medellín after separating from her husband, the father of her four children. "Like a mule!" she exclaimed when the subject of work arose. She had worked like a mule, she meant, all her life. Magaly explained that Aliria had washed and ironed people's clothes for a living.

The melanomas embarrassed Aliria, because they were ugly. They didn't hurt—except in her mouth, where they pushed through her gums—but they itched and caused fevers. The doctors with Neurociencias were very good to her, she said, very polite. She had had lots of fun with them in Boston, even though it was cold there. She had friends all over Barrio Pablo Escobar. Recently, she said, a neighbor had come to visit her who had recovered from coronavirus.

"See?" gushed Ana Baena, a clinician who, like Aguillón, had worked with Aliria for years and accompanied her on trips to Boston. "She remembers words like *coronavirus* and *Neurociencias*." Aliria was an extraordinary person, Baena continued. Not just because of her resistance to Alzheimer's but her attitude, her sense of humor. "She's always in good spirits, even in the face of adversity."

Aliria beamed on hearing that. She didn't like to complain, she said, and she would continue making people smile "until the last day. Until the day God takes me."

She was fascinated by a wallet card I carried of a saint, one that Blanca Nidia Pulgarín had given me, and which I ended up giving her. Magaly explained that her mother, when she was young, had suffered a wound that would not heal until she sought intercession from that saint.

Aliria fondled the card and arranged it by her pillow. She was growing tired from the attention and starting to fade, her blanket pulled up to her chin. We got up to say our goodbyes. Like a good paisa, Aliria admonished us for leaving so quickly and demanded we return soon. Then she nodded off.

A MONTH LATER on a cool, sunny morning came the inevitable message from the brain bank: no name, just news of a donation. **Is it her?** I wrote to Aguillón, who confirmed that it was. I had seen Aliria in one of her last good moments before her cancer began to exact its final toll and her organs started to fail.

Aguillón had gone up to Barrio Pablo Escobar himself, to oversee the retrieval of Aliria's body and get his papers signed—an unusual step, as he usually just met families at the funeral home. I parked myself in a coffee shop close to the SIU, not sure when the autopsy would get under way, and followed the WhatsApp traffic from the brain bank, where Andrés Villegas was issuing orders. Students and technicians were already in the lab preparing all the different preservatives they were now using, and checking the freezers.

At 11:30 a.m., Aliria's autopsy began. Most of these procedures were overseen by either Villegas or Aguillón, but this time they officiated together. "It's not every day we handle a case with these characteristics," Villegas told his staff solemnly as we all circled Aliria's body. "This is a historic moment. We're going to learn a lot from this case."

There had already been weeks of behind-the-scenes jockeying over who would get which parts of Aliria's brain when she died. Villegas had decided that Ken Kosik and the Germany-based researcher Diego Sepulveda-Falla would each be sent samples from brain regions of importance for their studies. The rest of the brain would remain in Colombia, where Villegas and his colleagues would conduct immunostaining for amyloid, tau, and other disease-linked proteins; advances at the brain bank now allowed much of this work to be done in-house. The Colombia team would also culture Aliria's astrocytes, brain cells that synthesize the APOE protein, and her melanomas. Tumor cells are basically immortal, Villegas explained, and if the scientists started a cell line based on Aliria—skin cells can be reprogrammed under lab conditions to become neurons—it would contain both E280A and two copies of the Christchurch mutation, allowing for all kinds of interesting studies. "Like HeLa cells?" asked Aguillón, referring to the tumor cell line from Henrietta Lacks. "That's right," Villegas said.

Aliria's melanomas had grown and proliferated since I'd seen her—or maybe it was that her clothes had concealed them on my visit. They pushed up like dark marbles under the skin of her exposed chest, and some had ulcerated.

It was my first time attending an autopsy since COVID, and everyone was wearing plastic hooded coveralls along with goggles, face shields, and N95 masks. Muñeton, the prosector, was using a different type of bone saw, one attached to a vacuum to trap infectious material. Within an hour Aliria's brain had been extracted, along with samples from several of her organs, and rushed to the lab at the SIU for dissection.

Villegas unwrapped his sterilized salmon knife and commenced cutting. To the researchers' relief, cancer did not appear to have reached Aliria's brain. The brain itself was small, but it was unclear whether this was due to atrophy or because Aliria herself was small.

Pinned to the wall was a map of a brain that stipulated which parts needed to be dissected in what order for electron microscopy and single-cell sequencing. Villegas had only ever followed this scheme a few times before; it was exacting and time-consuming, when time was of the essence. Adding to the pressure was that Villegas had so many samples to cut: eighteen regions to be preserved for Sepulveda-Falla, five sections to be frozen for Kosik, and more still for others. "If this were a normal brain we'd be done a long time ago," Villegas told me, wearily.

At 3:30 p.m., the dissection lab, no bigger than a walk-in closet, was feeling especially cramped. Jugs of preservatives, specimen cups, labels, and camera equipment crowded every surface as Villegas continued to cut, frequently stopping to squint up at the map. Trays and instruments were getting bloody. People bumped into one another. It was now seven hours since Aliria's death, and no one had eaten or taken a break. But the process was well advanced, with labeled parts going into the freezer and into preservative. Kosik's grad student called Aguillón with her checklist for the sections that would be sent to California by courier the next day, even as the pandemic limited flights out of Colombia.

I left while the team was still working, and spoke to Kosik that evening. Most of the day's efforts had been funded by the NIH grant he had spent years procuring, so he was surprised to hear how much of Aliria's brain tissue had been reserved for Sepulveda-Falla. When I spoke with him two days later, Sepulveda-Falla agreed that all the samples he'd requested had probably prolonged the dissection process, but he wasn't about to miss out: this famous brain was "a first come, first served" situation, he said.

Kosik recognized the importance of studying Aliria's brain, given the attention the case had generated. But he was not certain that it would lead to a major new finding. Sepulveda, by contrast, thought Aliria's autopsy brain might produce significant insights, especially when compared with results from other E280A brains. "This mutation shows all kinds of patterns of amyloid and tau," he explained. "You'd think that with brains of people from the same region with the same mutation and similar lives, similar diets, and a genetic background that's fairly uniform, you'd expect to have some leads. But I've seen a lot of variability."

Alzheimer's, Sepulveda-Falla continued, "is not a monolithic disease, even in this seemingly homogeneous population with the same mutation. If we're seeing this variability in a mutation population, what about the general population? A lot of the problem comes from the medical perspective. Physicians like to put everything in boxes, small clearly labeled boxes, but nature is not like that, and neither is disease for that matter. So I would say that I don't really know what

Alzheimer's is. And I have been studying this twenty years. Whatever it is, is a complex phenomenon."

TWO WEEKS AFTER Aliria's death, I went back up to Barrio Pablo Escobar to meet with her daughters, this time in the first-floor apartment where Aliria had lived. David Aguillón accompanied me again as a chaperone; even after procuring Aliria's brain, they still weren't taking chances with her family.

Aliria's sitting room was dimly lit and spare, with an old set of furniture and tall stacks of her CDs. Aliria had owned the home and shared it with her longtime companion, whom her daughters fondly called *el viejito*, the old man. El viejito was gone. He found the house so sad without Aliria that he left every morning looking for something to do.

The house sat right across the street from the notorious Escobar mural, which meant that Aliria had been greeted by Escobar's leering face whenever she opened her door. Rocío, Aliria's younger daughter, didn't give the mural much thought. "I guess it's some sort of cultural heritage thing," she said.

Rocío was fifty-four and glitzy, with ironed hair, shiny pants, and heels. She did not live in the barrio, and was driving a cherry red sedan with tinted windows. Her older sister, Magaly, dressed in sweats and seldom left her home just around the corner.

The women were in that early phase of mourning that is tinged with anger—I remembered it well from Daniela—and keen to assign blame for the death of their beloved mother. Could Aliria's cancer have been caused by radiotracers used by the scientists in Boston? Magaly demanded. Aguillón said that was doubtful. Maybe, Rocío conceded, her mother's obsessive sweeping of her stoop every morning had exposed her to too much sun.

The year before, when dozens of news articles hailed the discovery of a Colombian woman who'd beaten Alzheimer's, "everyone but us" had heard about it, Magaly said. "And then we saw Dr. Lopera asking for help for my mom," she said, referring to the article in *El Tiempo*. Why had he said that Aliria needed a nursing home? Was he planning to take her from them?

Aguillón pushed back politely on the complaints and accusations. Lopera had been misinformed about Aliria's situation when he made that comment to *El Tiempo*, he said, and no one had meant to keep them out of the loop. Magaly nodded and squinted, as though weighing what to believe.

Aliria's daughters had a strong rapport with Aguillón, whom they'd worked with a long time. But they had long ago discerned that the group needed them more than the other way around. For a while during the pandemic they'd stopped taking the team's calls, putting everyone on edge. They weren't exactly poor; in fact, Rocío seemed rather well off. But it was Rocío who had demanded, and won, a generous monthly subsidy for their mother. If Aliria was such an important patient, she felt, the scientists could pony up.

In the end, of course, Rocío and her sister had agreed to donate Aliria's brain and organs, signing a consent form that denied them compensation or royalties related to "discoveries, inventions or the development of a commercial product." The form may have been standard, but I wondered whether they would have gone ahead and signed it knowing a patent was in the works, or that millions of research dollars had already been awarded over Aliria.

Rocío said she was glad they'd donated their mother's brain, because while she didn't know much about Alzheimer's research, it was evident to her that things had stagnated. "I think the death of my mom, as sad as it is for us, may have opened many doors," she said, running her fingernails tenderly through her sister's fine hair.

"We're not selfish," Magaly affirmed. "We wanted to donate this brain. We hope the field can advance."

AT ALIRIA'S FUNERAL MASS the priest had done something priests seldom did in these parts: he spoke warmly of Aliria, describing her journey from Angostura and how hard she'd worked to give her family a better life. But the full story of that journey was quite a bit more complicated, her daughters explained. Even Aguillón, who had spent a lot of time with the family, didn't know all the details. It involved a knife, a failed murder plot, a brutal attack, and a flight from Angostura, all of which had occurred over the course of a single day nearly fifty years before.

In 1972, Magaly said, Aliria was a hardworking young *campesina* with four young children: two girls and two boys. Her husband, Gilberto, was a violent, temperamental man who moved the family around to other people's farms, or stashed them with his relatives while he found work elsewhere.

That year, though, Aliria had a stroke of luck. A local landowner was giving away small lots to poor families, and Aliria got one just for asking. "That was the greatest joy for us. I don't know where the money came from or how it happened," said Magaly, who was six at the time, but the family acquired materials to begin building their own home. "We skinned our shoulders carrying all those logs," she recalled, and lived with a neighbor while they built. "We had the house nearly ready, with only the door and the windows left to put in. Then one Sunday I see people removing the tiles and the logs. Imagine: my dad had sold the tiles, and the timber. I went inside and told my mom, 'Mom, Dad sold the house.'" Magaly recalled her shock and disbelief. "'Ma,' I told her. 'They broke it up. They broke it up, they're taking everything.'"

That same morning, Magaly remembered, a young man came by with a large knife, which Aliria calmly placed in her bag, and the young man left. "When my dad finished loading up all the things he'd sold, the old bastard, he went in and told my mom: 'I need to look in your bag,'" said Magaly. Gilberto opened the bag and discovered the knife. "And he grabbed my mom and took her outside, and smacked her across the hip with the broad side of his machete. Because she was keeping that knife to kill him with, he claimed."

Rocío interrupted her sister. "She *was* going to kill him."

"Well, that was the original plan," said Magaly. "But my aunt had already figured it out and talked her out of it, thank God."

Gilberto went on abusing Aliria in front of everyone. "He cut off a part of her ear and cut her face," said Magaly, and then, as though shocked by what he had done, dragged Aliria to a nearby creek to clean her wounds, as her children sobbed helplessly.

A relative intervened, forcing Gilberto to leave and allowing Aliria time to flee with the girls close behind. Aliria's youngest boy started after them, but he was too little to keep up, and Aliria couldn't risk turning back to grab him. "It breaks my heart even now to remember that moment," said Magaly. An injured Aliria and her daughters made it to the church rectory at Angostura, where the priest sheltered them overnight.

When they awoke the next day, the country radio station was broadcasting news of a woman on the lam with two girls: Gilberto had duped the announcers into painting Aliria as the villain.

Two days later, Aliria, Magaly, and Rocío reached a neighboring village, where a relative hid them in a farmhouse for weeks before they left, finally, for the city. Aliria carried away from the encounter a lasting injury: that first hard slap to her hip from Gilberto's machete had chipped her bone and caused an infection that suppurated long after Aliria and the girls had made it safely to Medellín.

Rocío turned to her sister. "Remember the smell? There are some smells that you never forget, and that one . . ."

"It smelled like grape soda," Magaly said. "I'll never forget it for the rest of my life."

DURING THEIR FIRST YEARS in Medellín they made five moves, from bad neighborhoods in the city center to bad neighborhoods on its periphery. Before they'd received their First Communions, Rocío and Magaly had seen armed robberies, apartment buildings ablaze, a dead man's feet poking out of a culvert. Aliria sometimes had no choice but to lock the girls inside while she went to work washing and ironing. It was in those precarious early years that Aliria met el viejito, a man she could trust with her girls and who helped keep everyone fed. Aliria's boys remained in the country, with Gilberto, and though they later came for periods to join their mother and sisters in the city, they found themselves more suited to rural life. The girls grew up hating the country, with its curses, envy, and relentless toil, with nothing ever changing and no way to progress. They mocked Aliria's country habits, like waking up before dawn to grind corn for breakfast.

"We never saw eye to eye. Never," Magaly said of Aliria. "But I loved her. I always knew and I always told her that no matter what, I would care for her in her old age," she reflected. "My mom was good people. Friendly, fun, a drinker . . ."

A drinker? I asked.

Magaly laughed. Aliria's binges were legendary. She would go on all-night benders with her girlfriends, and could often be spotted in billiard halls after midnight, drinking beer with one of her brothers. "I was thirsty," she retorted when her girls confronted her. Whenever she left the house with CDs in her purse, everyone knew she was about to go partying, and there was nothing they could do.

RESILIENCE

Even Magaly's late son, Dumar, who I gathered was a kind of gangster by the way Magaly talked about him—"he was the authority around here," she'd said casually—worried about Aliria's drinking. Magaly admonished her sons not to embarrass their grandmother. To each their own, she told them.

How long was Aliria like this? I asked.

All her life, until about three years ago, Magaly said.

I turned to Aguillón. Did he know this about Aliria?

He did, he said. And in the past year, when people speculated earnestly at meetings about "protective lifestyle factors" that might have helped Aliria beat E280A for so long, he had to keep from laughing.

Aliria was discovered by Neurociencias in 2015. It was generally held, prior to that, that people who did not get sick before a certain age could not be carriers of E280A, and healthy older people were not routinely tested. But in the final recruitment drive for the crenezumab trial, when the scientists were straining to get enough people enrolled, restrictions got loosened and more people were tested.

"In those days the police had captured Dumar," Magaly continued. "*Q'hubo* was writing about his case every day, so I got the paper every day." During that time the tabloid also published a front-page article on Lopera and his trial, which ran with a photo of the Piedrahita family. "I read the article and learned that they were bringing people in from Angostura and all those towns, and that there was a cousin of mine in charge of bringing them here," she said, referring to Ledy Piedrahita. Magaly learned of the group meetings in Medellín, health exams, blood draws, and genealogies. Everyone got money for taking part.

"I said to Rocío, 'Let's go,'" Magaly recalled, and they rounded up their brothers as well. Though Aliria's father—their grandfather—had "died stupid with Alzheimer's," Magaly said, they'd never viewed themselves as an Alzheimer's family in the same way as their cousins the Piedrahitas. They didn't have sick people in every room, the way the Piedrahitas did.

AT SEVENTY-THREE, IN 2015, Aliria had had her blood drawn. The person quickest to understand the importance of her surprise test result was Yakeel Quiroz, who started the paperwork for a visa to fly Aliria to Boston. Rocío accompanied her mother the first time, and again a few years later.

Rocío pulled up photos from those trips: Aliria in a pantsuit with red flowers, made up and smiling. Aliria in a wheelchair because she'd had knee surgery the month before. Aliria in a puffy jacket to protect her from the New England cold.

Rocío flicked through more photos and videos—many of them featuring Magaly's late son Dumar. There was a photo of Dumar on an expensive motorcycle; of a birthday visit to his well-appointed jail cell; of his funeral, in 2016, after he had escaped from prison and was killed. Magaly mentioned Dumar often as we talked, and I could see that her pain from his death remained raw. Before that she had been widowed twice, and now had lost her mother.

"Here's one!" chirped Rocío. It was a video of Aliria holding court in her hotel room in Boston, telling a story that had people laughing. She had lost memory toward the end, but Rocío and Magaly still could not bring themselves to believe that their mother ever truly had Alzheimer's. "You'd ask her how many children she had, and she'd get it wrong, but she still remembered people's names," Rocío said. Days before her death, when Aguillón had last checked in on Aliria, she'd greeted him by name. Aliria had remained lucid enough in her final months that she'd effectively signed off on her own brain donation.

Rocío, Magaly, and their brothers were all of interest to the researchers now. Each of the four carried a single copy of the Christchurch mutation. Which of them, if any, also carried E280A, only the scientists knew. Magaly and Rocío had made it into their middle fifties without problems, but their brother Albeiro, a farmer, now had memory lapses, they told me. He was fifty-two—a little late to begin developing symptoms, but not out of the usual range.

I asked Magaly and Rocío whether they would continue working with Neurociencias now that their mother had died. "We'll see," said Magaly, whose life had taught her never to play all her cards at once. "Depends."

TWENTY-SIX

By the spring of 2021, Daniela López Pineda's foot had healed and she was ready to work. But she had few good options in an economy struggling to recover from the pandemic. The first shipments of COVID vaccines had arrived in Colombia, bringing hope that the layoffs, lockdowns, and curfews might be coming to an end. But people continued to fall ill waiting for their shots.

The social and economic stress of the pandemic exacerbated political tensions, and massive, weekslong protests erupted in the cities of Bogotá and Cali. Cars and buildings burned; roadblocks went up; a famous statue of a conquistador was knocked off its base.

Daniela watched protest videos from all over the country with a righteous fury. Her upbringing in Comuna 13 had given her a taste of a social order enforced through cruelty, and a lifelong urge to topple it. She tied a bandana around her face and headed to downtown Medellín to march with its students and trade unionists. In the evenings, when neighborhoods reverberated with the sound of protest sympathizers banging pans out the windows, she bent my new frying pan with the force of her frustration.

The only job offers coming her way were bad ones. Her old company in Bogotá called asking if she would help them open up a satellite office in a small town. They would pay the minimum wage, which in 2021 was equivalent to only $252 a month. Daniela could not afford to move and set up house on those wages, so she'd taken a job washing pots at a Medellín restaurant for $8 a shift, so sore by the end of each that all she could do was lie in bed for hours, tapping on her phone.

One Friday in May, just before a Mother's Day weekend lockdown was to start, Daniela headed alone to the ossuary that held her mother's remains. She had something big on her mind, and wanted to talk and cry to Doralba, praying that a response would reach her from somewhere in the universe. Since the new year, Daniela had been thinking about emigrating to Spain. Women she knew from the comuna had done it, and returned to buy houses with their earnings. Many were sex workers, but Daniela had discerned from monitoring various groups on Facebook that there were other opportunities. "I'll work on a veal farm," she told me. "I'll pick olives. I'll can fish." You could do any of these jobs in Spain and, unlike in Colombia, not expect to be poor for the rest of your life. A big advantage of Spain for a Colombian, besides a common language, was that one did not need a visa to enter the country.

La Flaca could hardly stomach the thought of her sister abroad, but Daniela found an unlikely advocate in her brother-in-law, Fernay. She could expect to suffer, he warned her. Maybe for years. But at home, what could she expect beyond the minimum wage? Fernay was a hardworking, sober, and enterprising man, yet his best efforts were never enough to fund his modest dreams. He had recently moved his small gold-mining operation to a safer spot, but the threat of extortion remained. Even with a gold mine, you could still end up poor in Colombia.

I'd heard Daniela muse about many plans that never got realized. But when she talked about Spain, she sounded deadly serious. Daniela lacked the mule-like persistence to pull herself up step by incremental step, like her brother-in-law or her cousin Lina. What she did have was courage.

This Spain plan wasn't just about economics, she acknowledged. It was about Alzheimer's, too. She spoke frequently about her thirtieth birthday. She didn't talk about having a baby anymore, but still wanted a genetic test and hoped she could get one in Spain.

Most of all she seemed bent on upholding the vow she'd made in that consulting room with her sister and Andrés Villegas three years before: to make something of herself while she could. If she had fifteen years of good health left, those years would be hers. Daniela wanted to know how it felt to be, as she put it, "the absolute master of myself."

In late May, Daniela received her first passport. We went out for hot dogs to celebrate, just before the 8:00 p.m. COVID curfew kicked in. She took the passport with her, and flipped excitedly through its blank pages.

LOPERA, STILL WITH only one shot of COVID vaccine in his arm and awaiting another, spent his seventieth birthday at his finca with his immediate family, his orchids, and his dogs, fêted online by his Neurociencias staff. Many were still working from home, and most clinical and research visits—save for those related to the crenezumab trial—had been postponed, or were being conducted over the phone, somehow. The SIU's once-bustling courtyard was a ghostly place, with its cafés closed and access severely restricted.

In the Neurociencias offices, when I stopped by to pick up a document, only three people were working. Francisco Piedrahita, who could usually be found running from place to place with barely a moment to talk, was sitting quietly by himself.

I took the opportunity to ask him about Aliria's family, whom I'd recently gotten to know. It was curious to me that while Magaly,

Aliria, and Rocío were his cousins from the same hamlet, he never talked about them. And Magaly had spoken of Francisco's clan with cryptic reserve: "We never had much to do with them," she told me.

Just weeks before, Magaly made me lunch in her home, where she was recovering from a severe case of COVID. A blown-up photo of Aliria, against a celestial blue background, now hung over her dining room table. Lately Aliria had been appearing to her in dreams, Magaly said, as had her late son, Dumar, and her late husband. Sometimes she saw them all in the same room together: her beloved dead surrounding her, trying to say something.

Her sister Rocío turned out to be an independent businesswoman with a cattle farm, a trucking company, and a car wash she ran herself. Visiting her at the car wash, I found her to be a wonderfully diva-like boss, floating past the cars with her silky duster trailing her as she made for her office in the back. Years before, Rocío had worked her way up to senior management at a national fried chicken chain. When it was discovered that she had invented her diploma, the company debated firing her. They kept her, but she eventually struck out on her own.

Whatever this family was, they were not like the other Piedrahitas.

Francisco Piedrahita explained that he had known Rocío and Magaly all his life, and that yes, they were his cousins. He'd known Dumar, too, of course.

The reason his family and Aliria's didn't have much to do with one another, he said, was that his family was close with Gilberto, Aliria's estranged husband. He was Piedrahita's great-uncle, and had lived in their home for a time.

Piedrahita had had little contact with Aliria or her daughters in the years since they'd fled to Medellín. But he'd enjoyed being reunited with Aliria when she'd come into the Neurociencias fold as a star patient. He found her very lovable: "A sweet little lady who would get out the aguardiente and drink you under the table," he laughed.

Piedrahita himself was pondering a genetic test now that he'd turned forty. He would do it by petitioning his health insurance to cover it, he said. But the thought of receiving potentially life-changing results in such a quotidian fashion, "like a cholesterol test," gave him pause.

He followed the results of Alzheimer's drug trials carefully, including the back-and-forth over aducanumab, the anti-amyloid drug recently revived by Biogen. Piedrahita sounded as though he'd lost faith in the anti-amyloid approach, and I asked if him if he'd also lost hope for crenezumab. He allowed only that he was intrigued by the possibility of newer, non-amyloid-targeting drugs.

Every year, a group of scientists published a review of all the Alzheimer's drugs in the pipeline, and the most recent one had just come out, listing more than one hundred agents—an array that was surprisingly diverse in its targets and approaches. Some were aimed at stopping inflammation in the brain, or preserving the plasticity of neurons, or thwarting cell death. Others were designed to keep tau from misfolding or to improve the function of microglia, the immune cells of the brain.

It was incredible to think that another novel drug candidate might emerge from the study of Piedrahita's cousin Aliria. The Harvard team was working extensively with Christchurch, and also with a mutation on a gene called RELN that they considered responsible for the resistance of another man in the E280A cohort. This man had died in his seventies without much impairment, and brain pathology similar to Aliria's: lots of amyloid and little tau spread. Research groups elsewhere were exploring other protective mutations. A team in Florida was working with a mutation dubbed APOE3-Jax, for Jacksonville, that conferred strong resistance to late-onset Alzheimer's. Two research groups, one in Stanford, California, and another in Munich, were studying a variant on a gene called KL, or klotho, that seemed to keep APOE4 carriers from becoming ill. In a moment when the chances of a truly successful anti-amyloid drug were fast receding, the prospect that scientists might be able to replicate the activity of these "resistance" genes was a rare reason for hope.

ON JUNE 7, 2021, the FDA made the stunning decision to approve aducanumab for people with Alzheimer's disease. Days later, three members of the advisory committee that had voted resoundingly against the drug's approval resigned in protest.

There should have been much to celebrate with the approval of the first new Alzheimer's drug in eighteen years, but it only deepened the fissures in the Alzheimer's research world. The influential Alzheimer's Association, which aggressively lobbied the FDA in favor of aducanumab, was publicly shamed by Alzheimer's scientists who felt the group had become too enmeshed with drug manufacturers.

The FDA's decision stood to salvage the fortunes of not just the drug and its manufacturer, Biogen, but an entire drug class that had received billions in investments. Several more anti-amyloid antibodies—including the Roche drugs crenezumab and gantenerumab, Eli Lilly's donanemab, and Biogen's other antibody, lecanemab—were still in clinical trials. It seemed likely that, following aducanumab, one or more of these would also be approved.

Aducanumab's approval had been based on the controversial reasoning, upheld for years by advocates of the amyloid hypothesis and the drug makers, that lowering brain amyloid was "reasonably likely" to predict clinical benefit. The FDA was obliging Biogen to conduct further trials to confirm that the drug really worked, but gave the company nine years to complete the job. Until then Biogen had the right to sell all the aducanumab it could.

Perhaps the biggest shock came with the announcement of aducanumab's price. The antibody, to be branded as Aduhelm, would cost $56,000 per year of treatment, Biogen said—not counting any of the tests and scans. Health analysts began publishing scary calculations: if Medicare agreed to cover Aduhelm, it could cost the federal government as much as $29 billion per year, nearly double its entire current budget for prescription drugs.

Aduhelm was not the only wildly expensive drug on the market for a neurologic disease; some antibodies used in multiple sclerosis were priced in the same range, and new gene therapies for rare childhood disorders could cost in excess of $2 million. But Aduhelm was at best associated with a 22 percent slowing of cognitive decline—which equated to about five months of clinical benefit—and its full safety and efficacy results had never been published. The drug had already been seen in one study to cause brain swelling or bleeding in 40 percent of patients. The combination of sky-high pricing, questionable efficacy, and serious safety concerns turned Aduhelm's debut into a scandal that roiled well into 2022.

If Biogen could put such a price on Aduhelm, I wondered, what would crenezumab cost if approved? A single year of Aduhelm on the U.S. market was equivalent to eighteen years of the Colombian minimum wage. Prices for drugs were usually set much lower in developing countries. But even at $5,000 per year, it was difficult to imagine how a drug like this could become available to the E280A families, when patients sometimes had to sue their insurers to cover prescriptions that cost two hundred dollars. During periodic shortages, even cheap generic drugs could require legal maneuvers to obtain.

I WAS UNSURE HOW Lopera would feel about Aduhelm when I called him weeks later.

On one level the FDA's decision was bad, Lopera told me, because it had generated false expectations. A 22 percent slowing of cognitive decline "isn't a solution to Alzheimer's," he said, and the disastrously high price was making him worry about how Roche might price crenezumab. He hoped Aduhelm would not be approved in Colombia, because it would only lead to lawsuits, he said, and healthcare budgets would be drained.

And yet despite all these caveats, he continued, the FDA decision came with potential benefits for his research group. If crenezumab failed, one of the FDA-mandated aducanumab trials might be done at Neurociencias. If Lopera was to sustain the infrastructure and scientific production of Neurociencias he would need to conduct more clinical trials. In fact, without the crenezumab trial and all the investments and studies leading up to it, his team's contributions to Alzheimer's research would have been far fewer. The imaging findings, the biomarker discoveries, Aliria—these had all come about thanks to the trial and the investments it necessitated. But I still had trouble seeing new trials for aducanumab as a good idea for this population.

When the crenezumab trial was launched, nearly a decade before, its outcome was uncertain but there was reason for hope. For all anyone knew at the time, crenezumab could have been a knockout success, preventing or substantially slowing the development of Alzheimer's among E280A carriers. But the Aduhelm affair had shown how little could be expected from this drug class, and at what risk and price. To ask an E280A carrier, even a sick one, to try Aduhelm, knowing that at best it might slow the disease slightly and potentially cause a brain bleed in the process, and in the end it wouldn't matter because they could never afford it anyway: this was an entirely different proposition.

TWENTY-SEVEN

When Paula and I spoke by phone, in July 2021, she was calling to ask my help in getting a gift to her family in La Ceja.

Then, almost as an aside, she told me, "I have it." She meant the E280A mutation. She'd known for about a month.

I was shocked. She was not, she said. Ever since she'd learned about the mutation and the disease, nearly two years before, she'd assumed she was a carrier. She didn't know why she knew, but she knew.

Paula cried a little as she talked, but as always, she caught herself and regrouped.

She'd begun planning for the test before the pandemic, with a neurologist. She made her appointments privately, paying out of pocket. The neurologist referred her to a genetic counselor, who delivered the results over a forty-five-minute Zoom call.

Paula's mother, Miriam, came to see her immediately upon hearing the news. Paula had always wanted to return Miriam's generosity by caring for her when she got old. It broke her heart to think she wouldn't be able to.

Her husband, Sean, was trying to stay positive, Paula said, "but he's devastated." They were both concerned for their child, who was five. Paula was forty; by the time she developed symptoms, she figured, her son would be in his teens. But that was an ambitious calculation. Paula's biological mother, Mery, had developed symptoms in her late fifties, but her aunt Elida was in her forties when her symptoms began. Until the relevant resistance genes were better understood, precise age at onset wasn't something that could be predicted.

Only Paula's family and a few friends knew her results. She had told Marta, whose first reaction was to protest that it wasn't fair—they had only just reunited after having been apart for so long. Marta did not believe—she would never believe—that her sister had done the right thing by getting tested.

It bothered Paula to think she might have passed E280A on to her son. She wouldn't have had him, she said, had she known about all this beforehand. But she was glad she hadn't known, because he was the light of her life.

There was an inherent injustice in a mutation like this, something all the families felt and sometimes spoke about. In the past month, Paula had done her best to hold it together. She had dutifully gotten up every day, gone to work, and taken care of her family.

Still, she said, "I get angry. I get so angry. I had a terrible childhood, and now I get Alzheimer's to not really complete my life cycle. Not just Alzheimer's, but early Alzheimer's. I sometimes ask myself, what have I done to deserve this?"

I'D BEEN TRAVELING in the United States when I spoke with Paula. When I returned to my Medellín apartment, the next week, Daniela greeted me in tears. "I have a lot to tell you," she said.

Her uncle Fredy had spent the last months bedridden and unable to speak. Days went by when he did not sleep at all. The Neurociencias clinicians had already come to talk with the family about a brain donation. They prescribed him drugs to help him sleep, but he couldn't. He'd been up now for ten days. Daniela had just visited to find her beloved uncle shriveled and cadaverous, his green eyes staring at the ceiling and his mouth open in a way that reminded her of her mother.

Then Daniela revealed the news she'd been saving: Fredy's family had decided to let him die.

His wife and children were not feeding him or giving him water. Doctors with his health plan were coming daily to the house and administering sedatives and painkillers. Daniela had been shocked and angry when Fredy's wife first told her what was happening. But after speaking to her again, she felt differently. Fredy was calm, comfortable, and very close to the end. "He's on his journey," his wife had reported that morning.

This was the first time I'd ever heard of any of the families hastening a patient's death. There was talk of euthanasia, as there was of suicide, but as far as I was aware, no one had gone through with either. Euthanasia had been legal since 1997 in Colombia for terminally ill patients. But Alzheimer's fell into a gray area, in part because it was hard to tell how long a patient could expect to live, and because a patient's own desires could and did change once the illness set in.

Daniela sat on the bed in my guest room, which she'd made her own, amid a mountain of stuffed animals. Her high school portrait and photo of Doralba stared back at her from the door of the closet. In my absence, she had done a lot of planning for her trip. She'd chosen her flight, which she was hustling donations to help pay for. To persuade immigration authorities that she was a tourist, she'd created a fake itinerary that included tours of Toledo, train schedules, and hostels in different towns. She'd watched episodes of a National Geographic show about airport security, studying the questions the agents used to trip up Latinos entering Spain.

I told her—probably it was not the right time—about Paula's positive test result. She was shocked to hear it, and looked despondent, her face in her hands and her long hair hanging limply to her lap.

She was glad to be leaving. She needed a reprieve from all of this.

Of course her genes would travel with her; for Daniela, as for Paula, no geographical or cultural distance could stop E280A if she carried it. But she could at least avoid the repetition of blunt-force emotional trauma. Each illness or death in her family, each brain donation, each funeral—even news of a positive test result from someone she didn't know—reopened the same wounds. How many times could she go through this?

Fredy was not the last of her mother's generation to fall ill. It was increasingly evident that María Elena was impaired. Her facial expression had changed, and her hair looked wild. She was forgetting plans and posting strange religious messages on Facebook, claiming to be protected by the blood of Jesus.

FREDY'S BRAIN AUTOPSY was the last one I attended. In the years since my first, that of Fredy's sister Doralba, the brain bank protocols had changed beyond recognition. It had become a bustling operation, and the divides at the top—between Kosik's team and the Germany–Medellín–Harvard nexus—were no longer so bitter. Two new specialists had recently been hired in Medellín, a full-time brain anatomist and a pathology technician, and lots of machines had been purchased that fixed or cut or stained tissue with precision. All this was being paid for with Joseph Arboleda-Velasquez's grant money and Kosik's NIH funds. The brain bank's immunochemistry findings had been validated by a Banner pathologist in Arizona, showing that the Colombian results were as reliable as those of a top U.S. brain bank. Meanwhile, in Germany, Diego Sepulveda-Falla was carrying out the principal studies on Aliria's brain for a paper in progress, to which Kosik was contributing his single-cell sequencing results. Lopera was proud of having coaxed everyone to get along, at least for now, and the infusion of cash did not hurt. Medical students were increasingly eager to be part of the brain bank, vying for the chance to get called out of bed at three a.m. The bank was soon to log five hundred autopsy brains. Sometimes the team had to handle two donations in one day.

DANIELA PURCHASED A ROUND-TRIP ticket to Madrid by way of Amsterdam whose return she had no intention of making. A few days before her trip, she got an elephant tattooed on her inner forearm. Elephants were a symbol of Comuna 13, appearing frequently in the neighborhood's street murals memorializing the violence of the early 2000s. They were meant to represent the resilience of its residents after all that had happened to them—and their vow to never forget.

TWENTY-EIGHT

The API Colombia trial survived the pandemic, and by October 2021, the first participants to enroll had received their last scheduled doses of crenezumab. If Lopera was worried about a bad result, as many of his colleagues believed he was, he projected only optimism. In his emails to staff, he talked about Neurociencias's future in the brightest of terms: "Next year will be the most exciting in our history, as we will have demonstrated to the world that we can adequately evaluate the efficacy and safety of an investigational product, and then some! Regardless whether the molecule proves effective or not, a new era begins. . . ."

Lopera had yet to return to his offices at the university. The grounds of his finca had grown even lusher after a year and a half of pandemic lockdowns. Giant orchids sprouted from trees, and one bougainvillea was so thick and spreading that it threatened to topple a deck.

He was upbeat when I saw him there, surrounded by soft breezes and birdsong. The approval of aducanumab, and the possibilities for more anti-amyloid drugs, were exciting to him, even as Aduhelm had become synonymous with scandal in many corners of the Alzheimer's world. Lopera saw its approval as a partial ratification of the amyloid hypothesis, which he still had a little bit of faith in.

I asked him whether, if the amyloid hypothesis were to collapse, the Colombian families would become less relevant to science. The field had developed an amyloid-centric view of the disease in part because so many seminal findings had come from the early-onset families, whose genetic mutations caused an overproduction of amyloid. If these families, who represented just 1 percent of all Alzheimer's cases in the world, were no longer considered a good model for late-onset disease, would they be forgotten?

Lopera didn't think so. Both familial and late-onset Alzheimer's involved amyloidosis, tauopathy, and neuroinflammation, he pointed out. Moreover APOE, the key genetic factor in late-onset Alzheimer's, was at the heart of the Aliria discovery. Her case suggested that a mutation on APOE could stave off one of the world's most dreaded, aggressive forms of Alzheimer's for thirty years. "Thirty years," he repeated. "That's not a joke."

There would be—there had to be—more trials in Colombia. Lopera had been deepening his connections in the pharma-centric corners of the Alzheimer's world, including with the small cadre of researchers who loudly defended aducanumab and served firms like Biogen as speakers and consultants. Biogen had sponsored Neurociencias's latest Alzheimer's awareness campaign in Colombia, which was bigger, slicker, and drew more media attention than any before it. Ideas that had been previously viewed with caution, such as dosing in very young adults, and early-phase studies that could not confer clinical benefit, were being entertained.

If crenezumab failed in the E280A families, "we'll start again with another drug," Lopera said. His doors were open to partners old and new, and different types of trials. His vision was becoming more corporate. He liked the idea of offering the families a choice of drug trials to enroll in, as cancer patients might have. He would not have to send emissaries into the countryside, as he had done in years past, to fill trials. The families would come to him at a new complex to get their experimental treatments. At first blush the proposed complex sounded like the help center for the E280A families that people like Daniela had long hoped for. But what Lopera went on to describe was a high-end care facility, complete with a nursing home, stores, and hotel, for paying patients with Alzheimer's and other neurodegenerative diseases. The idea, which wasn't so different from that behind many institutions in the United States, was that funds from these patients would help support the group's research. Lopera was ardent about building his complex and working with the university to secure land for it in a wealthy outlying town. He wanted to name it after his most inspiring patient. *Villa Aliria* was the name rolling around in his mind.

WEEKS LATER I saw Aliria's daughters in Barrio Pablo Escobar for a Mass celebrating the first anniversary of Aliria's death. Magaly turned her home into a chapel for the occasion. She hung a crucifix, lit candles, arranged two dozen plastic chairs, and converted her dining room table into an altar. Rocío showed up early, with flowers to arrange. It was a dark, rainy day, with water sluicing violently down the steep alley outside.

Their brother Albeiro had come in from the countryside. A gentle, good-humored man with a farmer's wiry physique, Albeiro was fifty-four and appeared to be in the early stages of Alzheimer's disease. He spoke well but had a long gaze and sunken, shadowy eyes. His lapses were still minor, like forgetting to drink his juice with lunch. Like all of his siblings, Albeiro carried one copy of the Christchurch mutation. Did he also have E280A, and had Christchurch been keeping his symptoms at bay until now?

Magaly served a stew made from one of Albeiro's chickens. Albeiro, she said, always brought her chickens to cook. That made Magaly and Rocío remember something funny about their mother.

Aliria used to pet her chickens and make a big show of her tenderness before snapping their necks, they said. "See? This is why you can't trust anyone," she told them, laughing hysterically.

The rain paused long enough to allow Magaly's sons, various neighbors and relatives, and Aliria's longtime companion, el viejito, to arrive. The Mass started late and no one took communion. Afterward, everyone headed to the kitchen for coffee and pastries.

Magaly and Rocío were annoyed that the Neurociencias investigators hadn't remembered this day, the first anniversary of their mother's death. Magaly and Rocío felt they deserved special treatment, and this was hard to argue with. They didn't even know the full extent of it. They had not been told of Lopera's plans for a Villa Aliria, or United States Patent Application No. 20220227852, for a monoclonal antibody based on the presumed mechanisms of the Christchurch mutation.

Even when I eventually brought them up to speed on all these developments, they didn't know what to make of them. Rocío was far more concerned about getting her mother into *Guinness World Records*. As we drove together in the rainy darkness in her big red sedan, she told me she took part in studies "for humanity," not for Lopera or Neurociencias.

I HAD ASKED LOPERA, at his finca, what he thought he and his investigators *owed* the E280A families after all these years.

Lopera thought for a moment. "You're asking what is our obligation to the population?"

"That's right," I said.

"To care for them well," he said. "That is the objective."

"To make sure they feel supported, and not alone," added his wife, Claramónika, who oversaw the programs collectively called the *plan social*. Any drug company working with Neurociencias would have to continue funding that, Lopera said.

The researchers weren't the only ones responsible for the families' well-being, he insisted: "The responsibility belongs to the people who have the problem, to us the researchers, and to society." It was true that the Neurociencias staff could hardly stand in for the families' doctors or fix the structural issues that kept many of them poor. But one side of this partnership had prospered more than the other, and I wondered whether, if Lopera was becoming more businesslike in his thinking, the families might not also do the same. Otherwise, the best they could expect in the coming years was a little bit of help, and the chance to try medicines meant for others.

DANIELA'S EXCITEMENT AT having made it into Spain was short-lived. She enjoyed her first days in Madrid and sent pictures from her walks. She marveled at the languages she was hearing and the people she'd heard about all her life but never seen in the flesh: Asians, Africans, Roma. Days later she moved to a Latino neighborhood, to a group house of Colombians who had a connection to Comuna 13. From there things began to fall apart. Within weeks she came to realize that her roommates were overcharging her wildly for the space—and promising connections to fictitious jobs. Seasoned, unscrupulous migrants scammed the new migrants, and Daniela learned the hard way. She left the group house and found a bed to rent in an old woman's attic.

With her funds dwindling, Daniela printed her résumé and hit the streets. She quickly discovered that Colombians were not trusted as live-in maids or nannies, which was what many undocumented women did. She walked all over Madrid with her useless résumés. Her foot swelled from the cold and the walking, and she returned to her bed every night in pain. The old woman let Daniela fry batches of empanadas in her kitchen in the wee hours, when electricity was cheaper, and Daniela took them to construction sites to sell to migrant workers. But as the weather grew colder there were fewer men working. One of Daniela's motivations in leaving Colombia was to change her life before her thirtieth birthday. I wondered if she would even get to celebrate that birthday in Spain.

TWENTY-NINE

The results arrived abruptly, ahead of schedule, in a 10:00 a.m. telephone press conference hosted by the Banner Alzheimer's Institute.

Eric Reiman did most of the talking. "Sadly," he began, crenezumab had not been shown to stem cognitive decline in a statistically meaningful way. The results seemed to point in the right direction—the antibody appeared to slow the disease slightly compared with placebo—but the difference did not reach statistical significance. The numbers weren't robust enough to prove the trend was real.

API Colombia, Reiman continued, had proven that a long trial recruiting a "vulnerable population in a developing country" could be carried successfully to completion. The trial was "highly valued by the participants themselves," he said, something demonstrated by the fact that an impressive 94 percent of them had stuck it out to the end. Finally, Reiman addressed the E280A families directly: "We want you to know we are not going anywhere," he said.

There were no family members on that conference call. No notice had been given, and almost none spoke English anyway.

Pierre Tariot got on to describe some of the technical details of the trial. Lopera spoke only briefly to say he was glad the trial had been completed, even if he was disappointed by the result. When Rachelle Doody, of Roche, had her turn to speak, she praised the investigators for having enrolled people living "in a remote region of the world . . . even jungle regions." The trial had shown, Doody continued, that "if you're a patient who lives in a third-world country you can be part of this global effort to solve your problem."

And on it went.

Soon the reporters and the analysts started in with the kind of questions that pharmaceutical investors needed answers to: Did crenezumab show any effect on tau? What other biomarkers had changed? The topline results spoke only to the overall effectiveness of the medication and whether it met its endpoints. Only one journalist on the conference call, Pam Belluck of *The New York Times*, had a question about the E280A families, which she addressed to Lopera. How did they feel about the negative result?

Lopera responded by saying something vague about the high retention rate of the trial.

That was when I knew that the families had not yet been told. Reiman concluded the call, once again thanking the "courageous research participants" who'd made it all possible.

AT FOUR O'CLOCK that afternoon, Lopera sent the news out to the trial participants. His message, delivered by WhatsApp, was worded in a way that forced them to read between the lines.

> *Families: Crenezumab did not demonstrate clinical efficacy but we will continue with the medicine while its possible biological efficacy is evaluated.*
>
> *The initial analyses show that, though carriers who received medicine had better scores on cognitive tests, the*

differences did not reach statistical significance compared with those who had received placebo.

It was June 16, 2022. In six weeks, Lopera told participants, more definitive results would be released at a conference in San Diego, during the Alzheimer's Association annual meeting. After that he would invite everyone to the SIU, to tell them more.

Lopera's message left Marta Pulgarín Balbín confused—what was "possible biological efficacy"? Did crenezumab work or not? Marta's sister Paula, in the United States, had less trouble parsing the implications of the crenezumab news, which brought back that terrible "Why me?" feeling from when she first learned she was an E280A carrier.

Daniela, when I reached her in Spain, sounded as disillusioned as I expected.

"Now we're starting again from zero," she said. "The time has run out for the next generation. For me."

What about the other thing? Daniela wanted to know. The Aliria drug?

Nothing yet, I said. A scientist in San Francisco, Yadong Huang, had used gene editing to place the Christchurch mutation onto an APOE4 background in mice—moving it, in effect, from APOE3 to APOE4. The mutation kept the mice from developing Alzheimer's-type pathology, which was important. It meant that both early-onset Alzheimer's disease caused by presenilin mutations and late-onset disease in APOE4 carriers could potentially be targeted with a drug inspired by Christchurch. Creating that drug was a whole other story.

But there were other drug candidates, other approaches. Even the loudest skeptics in the Alzheimer's field, those who scoffed on Twitter at every anti-amyloid failure that confirmed their long-held convictions, still believed that major breakthroughs were inevitable. It was a question of when.

VALLEY OF FORGETTING

DANIELA HAD MADE IT through a harrowing winter in Spain. By that spring the soles were cleaving off her sneakers and all her makeup had run out; it surprised me to see her bare face on video calls. She was fortunate to land a bed in a women's shelter run by a Catholic charity, which moved her to the seaside city of Málaga. She lied to people back home that she was working and doing well, except, of course, to Flaca, with whom she spoke multiple times a day. Flaca was in the throes of planning her long-awaited dream wedding, which Daniela would be forced to miss. Her cousin Lina would take over as Flaca's maid of honor.

After feeling invisible for so long in a cold and bustling metropolis, it lifted Daniela's spirits to walk Málaga's sunny streets and chat randomly with fishermen. Social workers helped her initiate the arduous, multiyear process of becoming legal. Until then she worked under the table, by the hour, cleaning apartments, chopping vegetables, and caring for a boy with severe autism. If she was lucky she had fifty euro in her pocket by the end of each week. She celebrated her thirtieth birthday on the beach with French fries and a beer.

BY THE TIME the crenezumab results were announced, Aduhelm—far from the blockbuster Alzheimer's drug its advocates hoped it would be—had turned into a fiasco. Medicare refused to pay for it, Biogen had stopped marketing it, and the company's CEO had quit. News of yet another amyloid drug failing had many longtime observers of Alzheimer's drug research sounding exasperated if not downright cynical. "How long are we going to keep doing this?" wrote Derek Lowe, a columnist for *Science* magazine, on the crenezumab findings. "Always there's another trial, another agent, another approach coming, and that's where the hope resides, perpetually," Lowe continued.

Yet pharmaceutical firms were still betting heavily on the amyloid hypothesis. Lecanemab, a Biogen drug that appeared to perform slightly better than aducanumab, would go next before the FDA; within a year it would be approved.

IT SEEMED THAT no Alzheimer's Association meeting ever took place without being shadowed by fresh bad news or fresh scandal. In July 2022, just before things got going in San Diego, Charles Piller, an investigative reporter at *Science* magazine, published a story accusing Sylvain Lesné, a scientist at the University of Minnesota, of manipulating—or possibly fabricating—data that supported the amyloid hypothesis. Back in 2006, Lesné had claimed to discover a soluble, pre-plaque form of amyloid-beta, which he named Aß*56, that caused cognitive decline in mice. As it turned out, Aß*56 may never have existed, except as doctored images; other scientists had tried and failed to find it in humans. Lesné's work was hardly the pillar on which the amyloid hypothesis rested, but it had helped stoke enthusiasm for drugs targeting "toxic oligomers," pre-plaque forms of amyloid that were beginning to aggregate. This was what several of the anti-amyloid antibodies, including crenezumab, were designed to do.

The conference began on the kind of festive note it always did: with food, drink, and self-congratulatory speeches. The Alzheimer's Association executives gave little impression of anything amiss as its executives spoke of "moving in the right direction," a "golden age for neurodegenerative disease," and the need for "a complex solution that goes beyond one hypothesis." People continued to donate generously, leaving the association more flush than ever. It had spread around

$90 million in the previous year on dementia research, while its lobbyists helped secure billions more in government research funding.

I hadn't seen Lopera, or the rest of his team, since before the crenezumab news. He looked tired when I ran into him the next morning at the exhibit hall, where Genentech, the Roche subsidiary, was offering free cappuccinos with photo images printed on the foam. Lopera was slated to deliver a keynote speech that week, in addition to more results from the trial; the successful completion of API Colombia had secured his place as an Alzheimer's insider. The association had just poured a fresh $6 million into dementia research in Latin America, and Maria Carrillo, its chief science officer, credited Lopera for giving them the confidence to do so.

Kosik was no fan of this conference but had come to support Lopera. After the crenezumab results and Lopera's speech, Lopera and his colleagues would proceed to Kosik's home in Santa Barbara for a party. "It's not just about the science," Kosik said. It was about friendship. In some ways, Lopera and Kosik had grown further apart. Kosik had lost patience with a lot of the Alzheimer's world, while Lopera had many more entities and researchers competing for his attention. But with their lives and careers so fortuitously entwined for so long, they remained each others' close and trusted confidants.

EARLY THE NEXT MORNING, the API Colombia team presented its detailed breakdown on the crenezumab trial. Surprisingly few people showed up to learn the outcome of a trial that had attracted worldwide interest for years.

Lopera started off with a brief history of the E280A families. Before he got into his slides, though, he did something I'd seen many researchers do. He flashed his disclosures, or list of companies from which he received money, for what felt like a nanosecond, not enough time for the audience to digest. This was a common trick, but I wasn't sure where Lopera had learned to do it. He was now a paid speaker for Roche and Biogen.

Virtually all Alzheimer's researchers had some commercial ties. Some observers considered this part of why they were slow to criticize the anti-amyloid approach: they worked closely with companies invested in it.

Pierre Tariot flicked through his own disclosures with practiced speed before getting into the nitty gritty of the trial. Eric Reiman then broke down some of the biomarker results, thanked API's "heroic" participants, and made mysterious reference to "our next trial in Colombia."

I asked Tariot how he felt. "I'm at peace," he said. "This was a moonshot, and we got them back safely. We did no harm."

JASON KARLAWISH, THE RESEARCHER who had helped design the API Colombia trial, was tweeting up a storm at the conference. No one wanted to talk about Aduhelm, he wrote, and the crenezumab results were "maddening." When I saw Karlawish I asked him whether he thought anti-amyloid drugs should be tried any more in Colombia.

The prospect was less appetizing than before, he conceded. "But the alternative is to pack up and go."

I asked him whether he thought the crenezumab trial had been worth it. He said he thought it was. The E280A families had been "transformed" by having taken part in modern research, he said. They had become actors in a host of activities: being assessed and tested, talking with scientists, coming to understand dementia as a phenomenon observed and anticipated using the tools of science. "All of this is an experience," he said.

This was true. And the experience went beyond the science. The

API participants, and their families, had enjoyed a decade of caring attention from the trial doctors and nurses. People had been taken far out of their daily lives, flown to places like Phoenix and Boston. They had been exposed to journalists and news crews. A few had come to follow the scientific literature on dementia prevention and took measures to try to buy themselves extra disease-free years. If more members of the families went on to learn their genetic status, as they would soon be able to, that would be life-changing, too.

Of course now they also counted the experience of participating in a trial whose results were negative. Neurociencias staff had become concerned that younger generations of the E280A families were becoming reluctant to participate in research, having seen their parents not much helped by it. The staff had trouble getting them into the clinic for evaluations. Sometimes they'd say flippant things, like "call me when you have the cure."

WITH EVENTS LIKE the press call or the Alzheimer's Association conference, his foreign colleagues took the lead. But when it came time to face the families, as always, Lopera was on his own.

The event he held was like one of the prepandemic Christmas parties, starting in the morning in the auditorium of the SIU. Food tents were set up, along with a stage for a band. The staff, including Lopera, all wore black polo shirts with the Neurociencias logo. Together they formed a receiving line for the participants, in an unusual show of formality, as the auditorium filled. Marta and Marcela came in from La Ceja. Blanca Nidia did not make it. She was finished with the whole trial business, she said, glad to have done her part.

Lopera addressed the families as calmly and confidently as ever, this time cutting to the chase. Crenezumab was very safe, he said. But it was not very effective, and Roche had made the decision not to take the drug forward. Lopera thanked the families for their responsibility and commitment. "We will evaluate new alternatives," Lopera told them, and showed a slide listing other anti-amyloid drugs. "It's possible one of these will be selected for the families in Colombia," he said.

The sickest people in the cohort hadn't come, of course, and would never see another trial. Most of the people in the room were probably too old to be eligible for one either, though their siblings or children might. I just hoped a better class of drugs would emerge before the next one took place.

Lopera spoke that day, as he increasingly did, about Aliria. I knew Lopera was still focused on his Villa Aliria project. He talked about it all the time, and now, into his seventies and fresh from the pain of the crenezumab results, he seemed concerned about what his legacy would be, and keen to build something tangible. Aliria was from among the E280A families—she was one of you, Lopera reminded his audience. The fearless, funny, hard-partying *montañera* who had died two years before had emerged as a symbol of hope to rival the gypsies of Macondo.

The researchers played a video spanning a decade of memories of the trial, with loving commentary from the team. It showed the families and researchers gathered at parties, by Lopera's pool, in Pierre Tariot's dining room in Phoenix. It also featured images of the families' private events, people posing together with balloons and birthday cakes against the rough brick walls of their homes. I saw Marta Pulgarín Balbín crying as the video ended.

She had struggled emotionally after the "downer," she said, of learning that crenezumab had failed. Her voice and affect seemed disembodied, her blue eyes distant. She had been dreaming, she told me, about people in her family who had died of Alzheimer's: Her two aunts, Oliva and Elida, and an uncle who had died before them. She dreamed of her late grandmother and the grandfather she'd never met, dancing on the steps of the church in Yarumal, her grandfather in a hat and brown suit.

Would she seek her genetic result? Lopera talked about genetic counseling encouragingly now, without mentioning suicide. He urged people to take advantage of the program that was just getting off the ground, and to talk to their families about taking part. Even if you don't choose to learn your personal status you will learn a lot, Lopera promised; it will be a valuable experience.

Marta said that under no circumstances would she seek her results. Marcela wouldn't either. They were convinced that their sister Paula had harmed herself by learning hers, though I disagreed. Paula had her moments of despair, to be sure. But she had also changed jobs to have more time with her son. She had canceled plans to adopt a child that she might have been too ill to care for. She had even called the DIAN researchers about taking part in a trial, although ultimately she decided against it: the anti-amyloid drug being used didn't appeal to her.

Marta and Marcela both said they would participate in future trials if invited. But they would do so with reduced expectations. "Our experience at first was so full of hope, that the medication would be effective, that we wouldn't end up like our aunts and uncles," Marta said. Her children, now in their twenties and thirties, had no interest in taking part in studies.

"Where's Pacho?" someone yelled. Francisco Piedrahita wasn't there. I wondered what had caused him to miss this important day. Later he told me: He was too sad. He couldn't bear to come.

As the day wore on the mood lifted. The university band played.

Waiters served a country-style lunch. No one acted in a hurry to leave. Lopera danced and danced with the trial participants. He danced to so many songs with so many partners that he was dripping with sweat by the late afternoon. Had any other clinical trial, anywhere in the world, ended in such a way?

THIRTY

When I first met Ken Kosik, in 2017, his colleague Francisco Lopera was already one of Colombia's best-known scientists, respected at home and abroad. The crenezumab trial had been successfully launched, following decades of fieldwork in near-impossible conditions. And yet Kosik continued to grapple with a perception, among U.S. Alzheimer's researchers, that nothing truly significant had come out of the Colombia project.

"You know how scientists are," he told me at the time. "Everybody worries about these petty little things, and you could easily say, 'What has the Colombia story really contributed to Alzheimer's knowledge?' In the history of the major breakthroughs, it's not like there's been any one big thing." But he had a feeling that this was soon to change.

Nearly a decade later, no one dismissed Colombia. Lopera, who once accepted honors from the city council and his hometown, was racking up high-profile international awards for the totality of the effort to which he and several of his colleagues had dedicated their lives. The fact that the API Colombia trial had not resulted in a licensed treatment, or a verdict on the amyloid hypothesis, was beside the point.

The Colombia project proved in the 1990s that presenilin was the sought-after early Alzheimer's gene. It produced the first insights into the buildup of amyloid plaques in presenilin mutation carriers, and revealed the subtle brain and cognitive changes present in them long before symptoms appeared. It had given rise to a type of clinical trial that had never before taken place, one for which there was no template. It had revealed Latin American genomes to be a rich source of dementia-causing, and possibly dementia-preventing, variants, and helped underscore the importance of ancestry in how these genes are expressed. It marked a violence-weary nation once ruled by drug capos as an exporter of science, from its own institutions and through its scientific diaspora. And while nothing resembling an Alzheimer's cure was on the horizon, despite decades of research and billions of dollars, it had shone faint light on a pathway toward one.

Genetic resilience to Alzheimer's—seen so vividly in Aliria and the other E280A outliers—was only beginning to be understood. It remained unclear, despite the Harvard–Colombia team's bold and frequent claims, whether they'd conclusively proven their case for RELN or APOE3 Christchurch. Like the longtime critics of the amyloid hypothesis, there were scientists who continued to doubt whether such variants could blunt the effects of E280A on their own. Yet regardless of how strong the evidence was for any specific genetic variant, the phenomenon of resilience was real. It was being documented most extensively in cases from Colombia, and it was the future of Alzheimer's research.

Four years after Aliria's death, her brain continued to produce fresh insights. Diego Sepulveda-Falla's work revealed that parts of the brain normally most vulnerable to tau pathology had been spared in Aliria, and that tau behaved differently in her tissues. Kosik's single-cell sequencing results showed that in Aliria's brain, tau got shunted into astrocytes, the cells that supply lipids to neurons. In other words, "the tau is diverted to a different cell type where it's rendered harmless," he explained. The same thing didn't occur in other E280A brains, nor in the brains of people who had died with late-onset disease.

It was not just Aliria's brain contributing to findings like these, but all the brains, all the blood, all the painful and intimate conversations in the bare-bones consulting rooms of a Latin American public university. It was the midnight meetings in funeral homes and the countless cups of aguapanela in remote country homes. The generosity of the Colombian families and the persistence of the investigators had produced a stream of knowledge that had become, in recent years, a torrent.

Sadly, by then it had become clear that Lopera would not live to see any breakthrough therapies that stemmed from his decades of work. In the summer of 2024, he disclosed that he had been diagnosed with metastatic melanoma, the same disease that had claimed the life of Aliria, and was undergoing intensive treatment. By the time the families were made aware of his condition, he was too ill to deliver the news himself. On September 10 of that year, he died, and David Aguillón was named the new director of Neurociencias.

Would Lopera's work help Daniela's generation? I wasn't sure. Basic science in Alzheimer's disease was unveiling many promising and previously unknown pathways, but it took forever to translate such knowledge into a tested, approved therapy. Because the amyloid hypothesis had resulted in licensed treatments—albeit ones that worked only slightly and temporarily, and with worrisome side effects—amyloid was still viewed by drug companies as a target worthy of further investment. Less than two years after the disappointing end of API Colombia, the Banner Alzheimer's Institute and Neurociencias began testing another experimental anti-amyloid drug, again one made by Roche, in E280A carriers as young as eighteen. This medication, called a gamma-secretase modulator, was different from the antibodies, but the idea was still that targeting amyloid early would help. A new API trial was in the works, backed by nearly 75 million dollars in NIH funding, that would test this drug alongside one of the approved anti-amyloid antibodies in two hundred people from the E280A cohort. Was this the best that could be done for the families? Or was it a way to keep the research gears turning and money and data flowing with so few alternatives available?

At Flaca's wedding, her mother's generation was absent; the guests older than fifty came from the groom's side. María Elena did not attend. She had become palpably impaired by then, unable to send a coherent text message or even make herself lunch. Perhaps she'd seen news about Lopera and gotten confused, or perhaps she conjured the idea out of her own desperation. But María Elena became firmly convinced that an Alzheimer's vaccine was ready, that Lopera had finally done it. She phoned her nieces and nephews, elated, urging them to accompany her to Neurociencias. The next generation already knew that this couldn't be, that things were vastly more complicated. This was what the families had learned over forty years of taking part in science: to resist undue hope and to resist despair.

They harbored other types of knowledge about Alzheimer's disease, which they carried, with no formal way to catalogue or transmit it, along with their coveted genes and biomarkers. That knowledge came from their repeated exposure to the disease's every stage, from their intimate bonds with its victims and from their fears for themselves. Daniela, who continued to struggle in Málaga, still without papers or stable work, took pains to document her recollections from Doralba's last months. She tried to understand how her mother had perceived the world in the state she was in. She sent me an image of some handwriting in a notebook, the results of her informal, retrospective, observational study.

Her expression was sometimes one of astonishment. She opened her eyes and made a sound and finally she smiled. She didn't remember me anymore in her mind. Even as she tried to do it, she couldn't locate me. But she had me in her soul and in her heart because she could feel me, and although she couldn't say it, her eyes could. They spoke, they turned greener, very intense and clear, accompanied by tears. Her fight was to maintain that feeling. That was the fight she had inside her, trying to keep that fire from turning to ash. Love is not remembered but felt.

ACKNOWLEDGMENTS

My first debt is to Ken Kosik, whom I emailed one Sunday night in 2016 after seeing him in a *60 Minutes* story about Francisco Lopera and the Colombian Alzheimer's families. He immediately encouraged me to undertake a book project, something he had long meant to do but had never found the time for. Ken shared his travel diaries and notes and, over seven years of chats and calls, he helped me parse the biological and moral stakes of the effort to which he had given so much of his life—not just the research in Colombia, but helping the rest of the world appreciate the sacrifices and achievements of his friend Francisco Lopera.

Francisco Lopera, besides being generous with his time and his spirit, allowed me to do what no reporter before had done: work unfettered. I got to know him, his research team, and their study participants on their own terms, over a long period. This represented no minor challenge or risk for him and his group, yet at many points over the course of this project, when I confronted some obstacle, he graciously intervened on my behalf. Without his transparency, authority, and respect for journalism, none of this would have been possible. The period I spent with Lopera would turn out to be his final years, yet they were as productive and exciting as any during his extraordinary life.

ACKNOWLEDGMENTS

There are twenty-five known branches, with nearly six thousand members, of the E280A families. During my time in Medellín, more families—including those with Alzheimer's variants besides E280A—came into the fold. All count important struggles and triumphs and histories, of which I got to know but a scant handful. The López Pinedas, especially Daniela, Flaca, and Lina, were the family who let me deepest into their hearts, their past, and their world. I am deeply grateful to them and to Marta Pulgarín Balbín, her sisters Marcela and Paula; Piedad Rua and her nieces; Blanca Nidia Pulgarín and her family; Magalys and Rocío Villegas; and others who shared their experiences with me, if even for an hour. And there were still more who welcomed me into their homes and provided me with beautiful and moving interviews but whose stories, sadly, ended up being cut from this manuscript, notably Hernando Pulgarín and family; Carlos Piedrahita; and the Cuartas family: Darío, Adriana, Isabel, and Rosa.

While I received help from nearly everyone at the Grupo de Neurociencias de Antioquia, and many of its collaborators in years past, I especially thank Sonia Moreno for her insights, her enduring friendship, and her passionate dedication to her patients with Huntington's disease, an important population that is not discussed in this book. Special thanks also to Margarita Giraldo, Juliana Acosta-Uribe, Claramónika Uribe, Amanda Saldarriaga, and Francisco Piedrahita. I thank Diego Sepulveda-Falla for our fruitful conversations, and Marisol Londoño and Ernesto Luna for helping me navigate the university ethics committee, to which I owe a debt on solving the tricky issue of surnames. David Aguillón and Andrés Villegas gave me access to the brain bank, while Eric Reiman and Pierre Tariot hosted my visit to the Banner Alzheimer's Institute in 2018, with Tariot answering years of follow-up questions. Lopera's colleagues Rafael Palomino and Alfredo Ardila provided me important context on Lopera's early career, though neither was named in this manuscript. Liliana Cadavid wel-

ACKNOWLEDGMENTS

comed me into her Texas home to tell the harrowing and heroic story of her research in the hill towns, which, I regret, ended up cut from the final version.

This project was hatched in 2017 in a small writing group in Daytona Beach, Florida, where I workshopped it with member writers Derek Catron, Jeff Boyle, Ginger Pinholster, and the late Sandy Smith-Hutchins. We were an unusually productive group, with good chemistry, who cheered one another on to produce several full manuscripts and published books in the five years we were together.

My clever coconspirator and agent, Lynn Johnston, had the foresight to get my book proposal into the hands of Cal Morgan at Riverhead. Cal helped me think about this book as it was evolving and refocus it time and again. His wisdom, humanity, and concern for people with dementia guided this editorial process. After Cal's departure in 2023, Courtney Young magnanimously took on the remainder of the project, which benefited further from her scientific knowledge and judicious editing. For eight years, Lynn, Cal, and Courtney kept this project moving forward.

Thanks to Patricia Clark for her careful legal review and interest in the topic; to Ariel So and Catalina Trigo of Riverhead who helped carry this along; and to Ryan Boyle, Janine Barlow, Cassie Mueller, Denise Boyd, Caitlin Noonan, Glory Anne Plata, Michelle Waters, Nora Alice Demick, Jynne Dilling, Ashley Garland, Kitanna Hiromasa, and Geoff Kloske.

For most of the time that I reported and wrote this story, I was on my own, without grants or institutional support—just my family and my assigning editors at WebMD/*MDedge* news keeping me solvent. But toward the end, when the manuscript was almost done, I got an invaluable gift from Stanford University's McCoy Family Center for Ethics in Society, which hosts an annual manuscript workshop for a nonfiction writer and provides a stipend to aid that person in the final

ACKNOWLEDGMENTS

stretches of a book. Joan Berry, Ashlyn Jaeger, and Anne Newman of the McCoy Center put together a dream team that included writers, geneticists, Alzheimer's specialists, Colombia experts, and scholars of literature and of research ethics to weigh in. The May 2022 workshop, and the contacts that resulted from it, made a huge difference for me. Thanks to Allison McQueen, Benoît Monin, Blakey Vermeule, Diana Acosta Navas, Emily Higgs, Elizabeth Mormino, Héctor Hoyos, Javier Mejia, Kelly Ormond, Leif Wenar, Mike Greicius, Patrick Phillips, Rob Reich, and my sister, Elizabeth Smith, who participated as both a neuropsychologist and an ally. Thanks also to Duana Fullwiley, Anthony Wagner, Meghan Halley, and Robert Aronowitz of Stanford.

British Colombian journalist Sophie Foggin spent months fact-checking this narrative with thoroughness, integrity, and discretion. Working jointly with Sophie was like having a caring coauthor who never tired of the minutiae.

Lina Builes Moreno transcribed my interviews with precision and tact for years and took real interest in the plight of the families.

Veronica Aristizabal Quintero and Mariana Correa, graduate students in history, uncovered documents relating to the Girardota family, as did the historian of Colombia Mary Roldán.

I did not seek to photograph the subjects of this book as I was usually interviewing them in situations that were already somewhat delicate. The family of Alirio Villegas kindly allowed their image to be used in the jacket design of this edition. Although they do not appear in the text, they are members of the C2 clan in Canoas and research participants with similar life experiences to those of their relatives whose stories are highlighted, and their home evokes many others I have visited.

Michael Suh, who has the literary equivalent of perfect pitch, read through this manuscript more than once with great care.

Marcel Villamil and Tatiana Lozano accompanied me early in my

ACKNOWLEDGMENTS

reporting to the towns of Yarumal and Angostura, while Jaime Pulgarín took me on the back of his motorcycle to Belmira. These were my first trips in the hill towns, and it was so gratifying to share them with curious friends.

Another good friend, the Colombian Australian science writer Andrew Wight, became a wonderful sounding board on the long walks we used to take, as we chewed over this project and the countless other science stories we will surely someday write.

Ellen Kotlow and her family, my lifelong friends, lived this terrible disease as this book evolved, and their struggle informed it in different ways.

Raymond and JoAnn Smith, my parents, were beyond generous throughout this entire process, understanding of my long absences, and ready with love, support, feedback, and advice.

My husband, Seth Robbins, moved to Colombia with me in 2018 and took a difficult job there, a selfless act that allowed me to focus on this work with an intensity that I could never have sustained otherwise. He shared in this reporting and the friendships that resulted from it, and was my final editor, portrait photographer, and champion.

NOTES

CHAPTER 1

3 **named Pedro Julio Pulgarín:** Surnames in this book are a mix of true names and pseudonyms. For decades Lopera requested that the press, both Colombian and foreign, mask the surnames of the people he studied. The reason for this was a desire to avoid stigmatizing or bringing unwanted attention to the Alzheimer's families, many of whom share a handful of surnames. The custom began to erode in recent years as more families asked—and in some cases demanded—that their true surnames be published. For this book I offered each family the option to choose. I used only the given names of some subjects, where possible, as a way of de-emphasizing surnames. A small number of first names were altered, usually at the request of the individual. In very few instances an identifying detail was also changed. I do not indicate where these changes occur.

5 **"sent him home," Cornejo recalled:** Unless otherwise noted, all quotes in this text come from interviews. Among both the families and the researchers, many people were interviewed, in formal and informal circumstances, repeatedly over years.

8 **go to the scene:** Lopera showed a strong instinct for epidemiological detective work early on. When he was a young medical graduate working in the Caribbean coastal village of Acandí, two young brothers died of unknown causes in his hospital. Lopera took the unusual step of traveling to the family home, deep in the forest, where he found that the boys' surviving siblings had bites on their fingers from vampire bats. Suspecting rabies, he sent samples from the children's brains to a lab by boat, and when these were positive, he alerted the government. A rabies expert from Mexico was brought in to investigate, and Lopera joined him in capturing, studying, and destroying the infected bats.

9 **Colombian medical journal:** William Cornejo et al., "Descripción de una familia con demencia presenil tipo Alzheimer," *Acta Médica Colombiana* 12, no. 2 (March–April 1987): 55–61.

9 **in the American Midwest:** Leonard L. Heston, Dale L. W. Lowther, Carl M. Leventhal, "Alzheimer's Disease: A Family Study," *Archives of Neurology* 15, no. 3 (September 1966): 225–33.

NOTES

9 **another in Canada:** Linda E. Nee et al., "A Family with Histologically Confirmed Alzheimer's Disease," *Archives of Neurology* 40, no. 4 (1983): 203–8.

11 **memories and their sleep:** Gabriel García Márquez, *One Hundred Years of Solitude* (New York: Harper & Row, 1970).

16 **up around the city:** "El Plan Macabro del Paramilitarismo en Antioquia," 1987: Antioquia bajo el yugo paramilitar, *El Espectador* (2017).

17 **city of leftist influences:** "La muerte ronda la U," *Semana,* September 6, 1987.

17 **course at a time:** Héctor Abad Faciolince, *El olvido que seremos* (Barcelona: Seix Barral, 2007).

17 **active in left-wing politics:** "27 años de la desaparición forzada, tortura y asesinato de Francisco Eladio Gaviria Jaramillo, militante de la Unión Patriótica-UP, cuyo crimen fue declarado recientemente como de lesa humanidad," *Hijos Bogotá,* December 10, 2014.

17 **kidnapped briefly by guerillas:** Lopera's kidnapping occurred in Acandí, on Colombia's Caribbean coast. There, a FARC representative entered the local hospital and forced Lopera to follow him onto a small plane. After landing he was led on horseback into the jungle, at night, to a guerilla camp. Lopera was not equipped to operate on the injured fighter, who had been shot in the femur, so instead agreed to provide the guerillas with a letter giving the man an assumed name and attesting that he needed an operation. With Lopera's letter in hand, the fighter was able to cross the border into Panama undetected and be treated. The FARC sent Lopera a copy of a Marxist–Leninist book in appreciation, though Lopera was no supporter. Years later the group killed one of Lopera's uncles for failing to pay extortion fees.

CHAPTER 2

19 **"understand the question":** Konrad Maurer, Stephan Volk, and Hector Gerbaldo, "Auguste D and Alzheimer's Disease," *The Lancet* 349, no. 9064 (May 24, 1997): 1546–49.

20 **hospital in Munich:** Ralf Dahm, "Alzheimer's Discovery," *Current Biology* 16, no. 21 (November 7, 2006): 906–10.

20 **"the Cerebral Cortex":** There are different translations of the title of Alzheimer's original paper: Alois Alzheimer, "Über eine eigenartige Erkrankung Der Hirnrinde," *Allgemeine Zeitschrift für Psychiatrie und Psychisch-Gerichtliche Medizin* 64 (January 1907): 146–8. For most of this section I referred to the paper by Gabriele Cipriani et al., "Alzheimer and His Disease: A Brief History," *Neurological Sciences* 32, no. 2 (April 2011): 275–9.

22 **early and late forms:** Robert Katzman, "The Prevalence and Malignancy of Alzheimer Disease: A Major Killer," *Archives of Neurology* 33, no. 4 (April 1976): 217–8.

22 **on the Pulgaríns:** William Cornejo et al., "Descripción de una familia con demencia presenil tipo Alzheimer," *Acta Médica Colombiana* 12, no. 2 (March–April 1987): 55–61.

NOTES

22 **protein called amyloid-beta:** George G. Glenner and Caine W. Wong, "Alzheimer's Disease: Initial Report of the Purification and Characterization of a Novel Cerebrovascular Amyloid Protein," *Biochemical and Biophysical Research Communications* 120, no. 3 (May 16, 1984): 885–90.

22 **precursor protein, on chromosome 21:** Daniel A. Pollen, *Hannah's Heirs: The Quest for the Genetic Origins of Alzheimer's Disease* (New York: Oxford University Press, 1996).

22 **known as the amyloid hypothesis:** John A. Hardy and Gerald A. Higgins, "Alzheimer's Disease: The Amyloid Cascade Hypothesis," *Science* 256, no. 5054 (April 10, 1992): 184–5.

23 **palinopsia in a child:** Francisco Lopera et al., "Phénoménes de Palinopsie Chez un Enfant: Presentation d'un cas et revue de la littérature," *ANAE (Approche Neuropsychologique des Apprentissages Chez L'Enfant)* 1 (1989): 59–65.

24 **a study of exorcisms:** The resulting paper appears to be lost to posterity.

25 **the medical examiner:** Alma Guillermoprieto, "Letter from Medellín," *New Yorker*, April 22, 1991, 96–109.

28 **old Indian bones and gold:** The tradition of seeking and unearthing what Colombians call *guacas*, or the interred remains and treasures of Indigenous peoples, is very much alive, despite laws against it, and attracts discouragingly little criticism.

CHAPTER 3

34 **results in *Nature*:** Manuel Elkin Patarroyo, "Trial of Human Malaria Vaccine," *Nature* 336, no. 626 (December 15, 1988): 158–61.

34 **Reynolds, like Patarroyo:** Pablo Correa and Lisbeth Fog, "Jorge Reynolds, el hombre que no inventó el marcapasos," *El Espectador*, June 29, 2019, www.elespectador.com/ciencia/jorge-reynolds-el-hombre-que-no-invento-el-marcapasos-article-868489.

CHAPTER 4

37 **Katzman in 1979:** Robert Katzman, "The Prevalence and Malignancy of Alzheimer Disease: A Major Killer," *Archives of Neurology* 33, no. 4 (April 1976): 217–8.

37 **battling at all costs:** Jesse F. Ballenger, *Self, Senility, and Alzheimer's Disease in Modern America: A History* (Baltimore: Johns Hopkins University Press, 2006).

43 **into the realm of the supernatural:** Germán Castro Caycedo, *La bruja: Coca, política y demonio* (Bogotá, Colombia: Planeta, 1994), 81.

47 **point about the disease:** Francisco Lopera et al., "Demencia tipo Alzheimer con agregación familiar en Antioquia, Colombia," *Acta Neurológica Colombiana* 10, no. 4 (October 1994): 173–87.

50 **room for doubt:** Consult Pathology Report, Department of Pathology, Boston Children's Hospital, 1995.

NOTES

CHAPTER 5

52 **Huntington's disease gene:** Marcy E. MacDonald et al., "A Novel Gene Containing a Trinucleotide Repeat that Is Expanded and Unstable on Huntington's Disease Chromosomes," *Cell* 72, no. 6 (March 26, 1993): 971–83.

53 **variant on APP:** Alison Goate et al., "Segregation of a Missense Mutation in the Amyloid Precursor Protein Gene with Familial Alzheimer's Disease," *Nature* 349, no. 6311 (February 21, 1991): 704–6.

53 **found on chromosome 19:** Daniel A. Pollen, *Hannah's Heirs: The Quest for the Genetic Origins of Alzheimer's Disease* (New York: Oxford University Press, 1996).

55 **killings a year:** Alcaldía de Medellín, *Caracterización del homicidio en Medellín: Periodo 2012–2018* (Medellín: Municipio de Medellín, 2019): www.medellin.gov.co/es/wp-content/uploads/2022/10/caracterizacion-del-homicidio-en-medellin.pdf.

58 **gene across seven families:** R. Sherrington et al., "Cloning of a Gene Bearing Missense Mutations in Early-Onset Familial Alzheimer's Disease," *Nature* 375 (June 29, 1995): 754–60.

58 **to the Colombian kindred:** R. F. Clark et al., "The Structure of the Presenilin 1 (*S182*) Gene and Identification of Six Novel Mutations in Early Onset AD Families," *Nature Genetics* 11 (October 1, 1995): 219–22.

CHAPTER 6

61 **had a name: E280A:** Francisco Lopera et al., "Clinical Features of Early-Onset Alzheimer Disease in a Large Kindred with an E280A Presenilin-1 Mutation," *JAMA* 277, no. 10 (March 12, 1997): 793–9.

61 **the spot on the presenilin-1 gene:** A second presenilin gene, presenilin-2, was also discovered in 1995. Far fewer Alzheimer's-causing mutations have been documented on presenilin-2 than on presenilin-1.

64 **Lemere and Arango published a paper:** Cynthia Lemere et al., "The E280A Presenilin 1 Alzheimer Mutation Produces Increased Aß42 Deposition and Severe Cerebellar Pathology," *Nature Medicine* 2, no. 10 (October 1996): 1146–50.

64 **it was later confirmed:** Teresa Gómez-Isla et al., "The Impact of Different Presenilin 1 and Presenilin 2 Mutations on Amyloid Deposition, Neurofibrillary Changes and Neuronal Loss in the Familial Alzheimer's Disease Brain: Evidence for Other Phenotype-modifying Factors," *Brain* 122, no. 9 (September 1999): 1709–19.

65 **book on her struggle:** María Edilma Pineda V, *Peste de la memoria: Todo murió en ella, menos el amor* (Medellín: Litográficos de Dacolor EU, 2001).

66 **painting of Jesus in every house:** The Sacred Heart of Jesus is a commonly displayed religious image in Colombia.

67 **popular science magazine:** Kenneth Kosik, "The Fortune Teller," *The Sciences* 39, no. 4 (July–August 1999): 13–17.

NOTES

68 **Alzheimer's disease caused by E280A:** Lopera, "Clinical Features of Early-Onset Alzheimer Disease."

CHAPTER 7

77 **by right-wing paramilitaries:** Cadavid was not the only researcher on the team to face such an ordeal in that period. At least two kidnappings reportedly occurred among the Neurociencias staff and those working in the field with them.

80 ***"La peste de la memoria":*** Francisco Lopera Restrepo, "La peste de la memoria en Antioquia," *Legado del Saber: Contribuciones de la Universidad de Antioquia al conocimiento* (Medellín: Universidad de Antioquia, 2002).

CHAPTER 8

81 **plaques did not accumulate:** Dale Schenk et al., "Immunization with Amyloid-Beta Attenuates Alzheimer-disease-like Pathology in the PDAPP Mouse," *Nature* 400 (July 8, 1999): 173–7.

82 **with disastrous results:** Keonwoo Yi, "Passive Immunotherapy—A Viable Treatment for Alzheimer's Disease," *Psychiatria Danubina* 26, suppl. 1 (2014): 256–65.

82 **excessive cholesterol buildup:** Akira Endo, "A Historical Perspective on the Discovery of Statins," *Proceedings of the Japan Academy, Series B Physical and Biological Sciences* 86, no. 5 (2010): 484–93.

89 **that her data:** Lennie Pineda Bernal, "Análisis bioético de la investigación de la enfermedad de Huntington en el estado Zulia, Venezuela / A Bioethical Analysis of Huntington's Disease Research in Zulia State, Venezuela," *Revista Redbioética*/UNESCO, año 1, vol. 1, no. 2 (2010): 50–61.

91 **they'd just seen:** Mario A. Parra et al., "Visual Short-term Memory Binding Deficits in Familial Alzheimer's Disease," *Brain* 133, no. 9 (September 2010): 2702–13.

91 **those of their noncarrier relatives:** Yakeel T. Quiroz et al., "Hippocampal Hyperactivation in Presymptomatic Familial Alzheimer's Disease," *Annals of Neurology* 68, no. 6 (December 2010): 865–75.

CHAPTER 9

96 **an exhaustive modern-day study:** Luis Miguel Alvarez Mejía, *Biologia, uso y manejo del arboloco (*Montanoa quadrangularis*)* (Manizales, Colombia: Universidad de Caldas: 2003).

97 **several members of the Piedrahita clan:** Pam Belluck, "Alzheimer's Stalks a Colombian Family," *New York Times,* June 1, 2010. Members of the family featured in the article were not identified as having the surname Piedrahita, per the custom at the time of obscuring surnames.

99 **as age twenty-eight:** Adam S. Fleisher et al., "Florbetapir PET Analysis of Amyloid-β Deposition in the Presenilin 1 E280A Autosomal Dominant Alzheimer's Disease Kindred: A Cross-sectional Study," *Lancet Neurology* 11, no. 12 (December 2012): 1057–65.

NOTES

100 **"genetic testing is not the norm":** Jason Karlawish, "Ethical Tangles in Neurodegenerative Disease Research," presentation at Columbia University Medical Center, New York, NY, March 27, 2017.

101 **found to cause brain swelling:** Christopher Carlson et al., "Prevalence of Asymptomatic Vasogenic Edema in Pretreatment Alzheimer's Disease Study Cohorts from Phase 3 Trials of Semagacestat and Solanezumab," *Alzheimer's & Dementia* 7, no. 4 (July 2011): 396–401.

101 **on the operating table:** In the 1970s and 1980s, newly minted general physicians serving an año rural were expected to perform surgeries; this has since changed.

CHAPTER 10

107 **for a disease:** Lesley Stahl, "The Alzheimer's Laboratory," produced by Shari Finkelstein, *60 Minutes*, CBS, aired on November 27, 2016.

107 **doses could be safely used:** Heather Guthrie et al., "Safety, Tolerability, and Pharmacokinetics of Crenezumab in Patients with Mild-to-Moderate Alzheimer's Disease Treated with Escalating Doses for up to 133 Weeks," *Journal of Alzheimer's Disease* 76, no. 3 (2020): 967–79.

108 **before becoming plaques:** Sofia Giorgetti et al., "Targeting Amyloid Aggregation: An Overview of Strategies and Mechanisms," *International Journal of Molecular Sciences* 19, no. 9 (September 2018): 2677.

CHAPTER 13

124 **collaborator Diego Sepulveda-Falla:** People living outside Spanish-speaking countries sometimes opt to hyphenate their double surnames.

144 **studies in familial Alzheimer's disease:** Nicole S. McKay et al., "Neuroimaging within the Dominantly Inherited Alzheimer Network (DIAN): PET and MRI," *BioRxiv*, preprint, posted March 30, 2022.

144 **first cognitive symptoms:** Randall J. Bateman et al., "Clinical and Biomarker Changes in Dominantly Inherited Alzheimer's Disease," *New England Journal of Medicine* 367, no. 9: (August 30, 2012): 795–804.

148 **arrived in the first half:** Pedro José Madrid Garcés, "La Dimensión Politica de la Cultura: Un estudio de las manifestaciones culturales afro en el municipio de Girardota—Antioquia," master's thesis, *Universidad Nacional de Colombia, Facultad de Ciencias Humanas y Económicas,* Medellín, Colombia, 2016.

149 **linked to Europe:** Matthew A. Lalli et al., "Origin of the PSEN1 E280A Mutation Causing Early-onset Alzheimer's Disease," *Alzheimer's & Dementia* 10 (October 2014): S277–S283.

149 **rare European mutations thrived:** Mauricio Arcos-Burgos and Maximilian Muenke, "Genetics of Population Isolates," *Clinical Genetics* 61, no. 4 (April 2002): 233–47. This and related papers by a group of University of Antioquia geneticists, produced in the late 1990s and early 2000s, had the cumulative effect of painting the Antioquia population as overwhelmingly white and European derived, with some Indigenous ancestry but a negligible contribution of African genes. While many Latin American populations have espoused similar beliefs about their origins, scientific ratification of them is rare. The geneticists'

NOTES

findings were celebrated in the Colombian press at the time of their publication, but criticized as biased by Colombia's leading genetics researcher, Emilio Yunis, whose own studies had produced different results. The papers were also singled out for criticism by the anthropologist Peter Wade in his 2017 book, *Degrees of Mixture, Degrees of Freedom: Genomics, Multiculturalism and Race in Latin America*, from Duke University Press.

150 **compared samples from the family:** Juliana Acosta-Uribe et al., "A Neurodegenerative Disease Landscape of Rare Mutations in Colombia Due to Founder Effects," *Genome Medicine* 14, no. 27 (March 8, 2022): 27.

150 **the I416T mutation occurred:** Laura Ramirez Aguilar et al., "Genetic Origin of a Large Family with a Novel PSEN1 Mutation (Ile416Thr)," *Alzheimer's & Dementia* 15, no. 5 (2019): 709–19.

CHAPTER 14

156 **transdermal patches of rivastigmine:** The drug, approved in the United States in 1997, is a cholinesterase inhibitor, which can temporarily improve brain function in people with dementia by increasing levels of a neurotransmitter.

CHAPTER 15

169 **Colombia team published:** Pierre N. Tariot et al., "The Alzheimer's Prevention Initiative Autosomal-Dominant Alzheimer's Disease Trial: A Study of Crenezumab versus Placebo in Preclinical PSEN1 E280A Mutation Carriers to Evaluate Efficacy and Safety in the Treatment of Autosomal-dominant Alzheimer's Disease, Including a Placebo-treated Noncarrier Cohort," *Alzheimer's & Dementia* 4 (March 8, 2018): 150–60.

172 **dozens of other pathways noted:** John Hardy et al., "Pathways to Alzheimer's Disease," *Journal of Internal Medicine* 275, no. 3 (March 2014): 296–303.

172 **Critics of the amyloid hypothesis noted:** Judith R. Harrison and Michael J. Owen, "Alzheimer's Disease: The Amyloid Hypothesis on Trial," *British Journal of Psychiatry* 208, no. 1 (January 2016): 1–3.

CHAPTER 16

181 **ineffective or unsafe:** Konstantina G. Yiannopoulou et al., "Reasons for Failed Trials of Disease-Modifying Treatments for Alzheimer Disease and Their Contribution in Recent Research." *Biomedicines* 7, no. 4 (December 9, 2019): 97.

181 **without serious safety issues:** Susanne Ostrowitzki et al., "Evaluating the Safety and Efficacy of Crenezumab vs Placebo in Adults With Early Alzheimer Disease," *JAMA Neurology* 79, no. 11 (November 1, 2022): 1113–21.

CHAPTER 17

195 **antibody was ineffective:** Ben Hargreaves, "Roche Alzheimer's Drug Suffers Trial Failure," *BioPharma-Reporter*, January 30, 2019.

NOTES

196 **with his text:** Kenneth Kosik et al., "A Path Toward Understanding Neurodegeneration," *Science* 353, no. 6302 (August 26, 2016): 872–3.

197 **news of the canceled trials:** Dave Morgan, "Roche Pulls Plug on Two Phase 3 Trials of Crenezumab," *Alzforum*, January 31, 2019.

200 **trials of aducanumab:** Adam Feuerstein, "Biogen Halts Studies of Closely Watched Alzheimer's Drug, a Blow to Hopes for New Treatment," *STAT*, March 31, 2019.

200 **wiping out $17 billion of value:** Takashi Umekawa and Tamara Mathias, "Biogen Scraps Alzheimer Drug Trials, Wiping $17 Billion off Its Market Value," *Reuters*, March 21, 2019.

208 **new Colombian–German collaboration:** Diego Sepulveda-Falla et al., "The Colombian–German Network for Neurodegenerative Research: UndoAD," *The Lancet Neurology*, 18, no. 1 (January 2019): 29.

210 **make genetic disclosure available:** Lopera and his colleagues would not administer the process, but they worked with an established genetics practice in Medellín.

CHAPTER 18

218 *lulo* **in its soils:** the tangy, rich lulo fruit is used in juices in Colombia. In Ecuador and Costa Rica it is known as *naranjillo*.

CHAPTER 19

233 **lambasting top Alzheimer's researchers:** Sharon Begley, "The Maddening Saga of How an Alzheimer's 'Cabal' Thwarted Progress toward a Cure for Decades," *STAT*, June 25, 2019.

235 **mutation carriers as young as eighteen:** "DIAN-TU Primary Prevention," Step Up the Pace, Alzheimer's Association, alz.org/stepupthepace/DIANTU-PRIMARY.asp.

237 **different ancestral backgrounds:** Juliana Acosta-Uribe et al., "A Neurodegenerative Disease Landscape of Rare Mutations in Colombia Due to Founder Effects," *Genome Medicine* 14, no. 27 (March 8, 2022): 27.

237 **Argentina, Puerto Rico, and Mexico:** Since 2018 a group of DIAN researchers led by the Cuban-born neurologist Jorge Llibre-Guerra has done exceptional work finding and studying early-onset Alzheimer's kindreds across Latin America, looking at social, geographic, economic, and lifestyle factors as much as genes and biomarkers.

238 **the trial's ethics and cultural sensitivities committee:** This advisory group was set up to sort out issues around genetic disclosure before the API Colombia trial, and became largely dormant thereafter.

239 **passing the costs on to strapped healthcare systems:** Adriana Petryna, *When Experiments Travel: Clinical Trials and the Global Search for Human Subjects* (Princeton, NJ: Princeton University Press, 2009). This important book focuses on Brazil, which leads Latin American countries in the number and scope of clinical trials conducted. Brazil has recently taken steps to better protect the

NOTES

rights of its citizens who participate in clinical trials, including giving them the option to withdraw any of their biological samples from further research use at any time; ensuring that people who become sick from a study treatment are cared for at the trial sponsor's expense; and forcing sponsors to continue providing certain study drugs until they become available through public-health channels. Similar measures, at the time of this writing, did not exist in Colombia.

240 **neurofilament light chain, or NfL:** Yakeel T. Quiroz et al., "Plasma Neurofilament Light Chain in the Presenilin 1 E280A Autosomal Dominant Alzheimer's Disease Kindred: A Cross-sectional and Longitudinal Cohort Study," *Lancet Neurology* 19, no. 6 (June 2020): 513–21.

CHAPTER 20

253 **drug company Biogen announced:** "Biogen Plans Regulatory Filing for Aducanumab in Alzheimer's Disease Based on New Analysis of Larger Dataset from Phase 3 Studies," press release, Biogen, October 22, 2019.

CHAPTER 21

257 **front page of *The New York Times*:** Pam Belluck, "Why Didn't She Get Alzheimer's? The Answer Could Hold a Key to Fighting the Disease," *New York Times*, November 4, 2019.

257 ***Nature Medicine*, just ten paragraphs long:** Joseph F. Arboleda-Velasquez et al., "Resistance to Autosomal Dominant Alzheimer's Disease in an *APOE3* Christchurch Homozygote: A Case Report," *Nature Medicine* 25, no. 11 (November 2019): 1680–83.

257 **mutation called APOE3 Christchurch:** Mark Wardell et al., "Apolipoprotein E2-Christchurch (136 Arg----Ser). New Variant of Human Apolipoprotein E in a Patient with Type III Hyperlipoproteinemia," *Journal of Clinical Investigation* 80, no. 2 (September 1987): 483–90.

258 **person's risk for Alzheimer's:** Elizabeth H. Corder et al., "Gene Dose of Apolipoprotein E Type 4 Allele and the Risk of Alzheimer's Disease in Late Onset Families," *Science* 261, no. 5123 (August 13, 1993): 921–3.

258 **to nearly nil:** Eric M. Reiman et al., "Exceptionally Low Likelihood of Alzheimer's Dementia in APOE2 Homozygotes from a 5,000-person Neuropathological Study," *Nature Communications* 11, no. 1 (February 3, 2020): 667.

258 **regarded as neutral:** Wenyong Huang et al., "APOE Genotype, Family History of Dementia, and Alzheimer Disease Risk: A 6-year Follow-up Study," *Archives of Neurology* 61, no. 12 (December 2004): 1930–34.

259 **"both the illness and the cure":** Heidy Yohana Tamayo Ortiz, "¿Podrá este hallazgo Colombiano darle al Mundo la cura del Alzhéimer?," *El Tiempo,* November 6, 2019.

259 **to Henrietta Lacks:** Rebecca Skloot, *The Immortal Life of Henrietta Lacks* (New York: Crown, 2010).

260 **one copy had any protective effect:** Some years later Yakeel Quiroz and Joseph Arboleda-Velasquez, along with several coauthors including Reiman, Lopera and Kosik, published results from twenty-seven Christchurch single-copy carri-

NOTES

ers who also carried E280A. They reported that having a copy of APOE3 Christchurch delayed the onset of symptoms by four or more years compared to E280A carriers without Christchurch, in a seeming vindication of the hypothesis that Christchurch could stave off disease. However, a geneticist who had previously worked with their data found that they had revised old clinical records for their analysis and adjusted the original ages of onset upward in some individuals, casting doubt on the integrity of the results. See Yakeel Quiroz et al., "APOE Christchurch Heterozygosity and Autosomal Dominant Alzheimer's Disease," *New England Journal of Medicine* 390, no. 23 (June 20, 2024): 2156–64; and J. Nicholas Cochran et al., "APOE Christchurch Heterozygosity and Autosomal Dominant Alzheimer's Disease," *New England Journal of Medicine* 391, no. 17 (October 31, 2024): 1660–61.

CHAPTER 22

266 **press their case:** Robert Hoban, "Hope for Yarumal: The Case for a Plant-Based Economy," LinkedIn post, August 13, 2019.

CHAPTER 23

275 **his "peculiar disease":** Alois Alzheimer, "Über eine eigenartige Erkrankung Der Hirnrinde," *Allgemeine Zeitschrift für Psychiatrie und Psychisch-Gerichtliche Medizin* 64 (January 1907): 146–8.

275 **tangles as tau:** K. S. Kosik, C. L. Joachim, and D. J. Selkoe, "Microtubule-associated Protein Tau (Tau) Is a Major Antigenic Component of Paired Helical Filaments in Alzheimer Disease," *Proceedings of the National Academy of Sciences* 83, no. 11 (June 1986): 4044–48.

276 **death of neurons:** Yang Shi et al., "APOE4 Markedly Exacerbates Tau-mediated Neurodegeneration in a Mouse Model of Tauopathy," *Nature* 549 (September 20, 2017): 523–7.

276 **tau uptake and spread:** Jennifer N. Rauch et al., "LRP1 Is a Master Regulator of Tau Uptake and Spread," *Nature* 580 (April 1, 2020): 381–5.

CHAPTER 24

286 **plan for his succession:** The question of who would lead Neurociencias after Lopera was a fraught one, though both David Aguillón and the Harvard-based team of Yakeel Quiroz and Joseph Arboleda-Velasquez had a hand in nearly every ongoing study by 2022. In August of 2024, with Lopera incapacitated and receiving treatment for a terminal cancer, Aguillón was named director of the research group.

290 **"but seventy-seven times":** Matthew 18:21–35 (New American Bible, Revised Edition).

292 **one percent of what was being spent on the API trial:** I never saw documents establishing that the budget of the *plan social* was $150,000 per year. Rather, it was described to me as such by people familiar with it.

NOTES

292 **saw considerable earnings from consulting fees:** Fee reports are drawn from the U.S. government website Open Payments (https://openpaymentsdata.cms.gov), which collects and publishes information about financial relationships between doctors and drug or medical device companies.

CHAPTER 25

295 **on the strength of the Aliria case:** The organization gave Arboleda-Velasquez another million dollars the following year.

CHAPTER 26

315 **most recent one had just come out:** Jeffrey Cummings et al., "Alzheimer's Disease Drug Development Pipeline: 2021," *Alzheimer's & Dementia* 7, No. 1 (2022).

316 **APOE4 carriers from becoming ill:** Michael E. Belloy et al., "Association of Klotho-VS Heterozygosity with Risk of Alzheimer Disease in Individuals Who Carry APOE4," *JAMA Neurology* 77, no. 7 (July 1, 2020): 849–62; Julia Neitzel et al., "KL-VS Heterozygosity Is Associated with Lower Amyloid-dependent Tau Accumulation and Memory Impairment in Alzheimer's Disease," *Nature Communications* 12, no. 1 (December 2021).

316 **was publicly shamed:** This mostly occurred on Twitter, and most prominently by Lon Schneider, a researcher with the University of Southern California's Keck School of Medicine, and Madhav Thambisetty, a neurology researcher with the National Institute on Aging.

CHAPTER 28

327 **Alzheimer's and other neurodegenerative diseases:** While the Villa Aliria project was first proposed as a care center for paying patients, with few obvious benefits to the early-onset Alzheimer's families, this changed as the project evolved. In 2024, a Switzerland-based entity called Herencia, which was developing the center along with the University of Antioquia, sought to create novel funding mechanisms that would give the families access to its care facilities and higher-quality social support, including through an onsite employment agency.

CHAPTER 29

333 **a drug inspired by Christchurch:** In 2022, Arboleda-Velasquez cofounded a company, called Epoch Biotech, to develop drugs based on mutations linked to resistance cases.

335 **Lecanemab, a Biogen drug:** Biogen developed Lecanemab jointly with the Japanese pharmaceutical firm Eisai.

335 **fresh bad news or fresh scandal:** The annual Alzheimer's research scandals did not abate. In 2023 Marc Tessier-Lavigne, an Alzheimer's researcher who was president of Stanford University, was forced to retract several widely cited papers over image anomalies and statistical errors committed by coauthors. That July, he resigned from Stanford. The following year, Eliezer Masliah of the National

NOTES

Institute on Aging, renowned for his research in Parkinson's and Alzheimer's diseases, was fired after an investigation revealed papers containing doctored images. Some of that research had been used in drug development for Alzheimer's and Parkinson's treatments. Ivan Oransky and Adam Marcus, "Science Corrects Itself, Right? A Scandal at Stanford Says It Doesn't," *Scientific American*, August 1, 2023, www.scientificamerican.com/article/science-corrects-itself-right-a-scandal-at-stanford-says-it-doesnt; "Statement by NIH on Research Misconduct Findings," press release, National Institutes of Health, September 26, 2024, www.nih.gov/news-events/news-releases/statement-nih-research-misconduct-findings.

338 **extra disease-free years:** One young man in the E280A cohort, the son of two affected parents, described to me his punishing daily exercise ritual and scrupulous avoidance of sugar. A preliminary study by DIAN investigators, unpublished at the time of this writing, has suggested, hearteningly, that lifestyle factors including exercise may delay the age of disease onset among people with presenilin mutations.

340 **taking part in a trial:** Outside of Colombia, members of E280A families could enroll in DIAN studies.

CHAPTER 30

343 **high-profile international awards:** In 2024, Lopera received the Potamkin Prize, which honors researchers who "advance the diagnosis, management, and search for a cure" for Alzheimer's and other neurodegenerative diseases. Ken Kosik, Eric Reiman, and Randall Bateman were all prior recipients.

344 **proven their case for RELN or APOE3 Christchurch:** This is especially true of the Christchurch heterozygotes.

344 **spared in Aliria:** Diego Sepulveda-Falla et al., "Distinct Tau Neuropathology and Cellular Profiles of an APOE3 Christchurch Homozygote Protected against Autosomal Dominant Alzheimer's Dementia," *Acta Neuropathologica* 144, no. 3 (September 2022): 589–601. Sepulveda-Falla's initial findings suggested that only certain structures in Aliria's brain were affected by tau pathology. Later he determined, using more sensitive tests, that Aliria had early tau pathology throughout her brain; what is uncertain is whether it would have continued to progress had she lived.

344 **Kosik's single-cell sequencing results showed:** Maria Camila et. al., "Single-nucleus RNA Sequencing Demonstrates an Autosomal Dominant Alzheimer's Disease Profile and Possible Mechanisms of Disease Protection," *Neuron* 112, no. 11 (June 5, 2024): 1778–94.

345 **By the time the families were made:** Articles celebrating Lopera's life and work appeared in all the major Colombian newspapers after he died, as well as *The New York Times*, *The Economist*, and *Nature*. His brain was donated to the Neurociencias brain bank.

346 **a new API trial was in the works:** Robert C. Alexander et al., "Alzheimer's Prevention Initiative ADAD Colombia Trial Program," NIH RePORTER, August 26, 2024, https://reporter.nih.gov/project-details/10855439.

INDEX

Abad Gómez, Héctor, 16, 17–18, 111, 146
Acandí, 356n
acetylcholine, 80
Acosta-Uribe, Juliana, 150, 185, 236–37, 266
aducanumab
  Biogen clinical trial, 200–201, 233, 253–54, 267, 315
  FDA approval, 316–18, 325
Aduhelm, 317–18, 325, 334–35, 337
African descent, 147–51, 185, 189–91, 236–37, 360–61n
aging, concept of, 37
agriculture and farming, 6, 8, 222, 283–84
aguardiente, 74, 199, 314
Aguillón, David
  background of, 184
  COVID-19 and, 286–87
  Neurociencias brain bank, 184–85, 262, 263, 273, 277, 286–87
    Aliria, 295, 296, 298–99, 301–4, 307, 308
    Camila, 202, 262, 263
    Elcy, 273
    Mabilia, 277–79
  named director of, 345
alanine, 61
Albeiro, 309, 327–28
  author's visit, 296–98
  background of, 244–45, 304–6
  brain autopsy and study, 299–302, 303, 322–23, 344–45
  death of, 298–99, 302, 327–28

E280A and APOE3 Christchurch mutation, 243–45, 254, 257–62, 267, 276, 289, 295–96, 308–9, 313–14, 315–16, 318, 327–28, 344–45
  funeral mass of, 304, 327–28
  Lopera and, 259, 261, 267, 303, 313–14, 327, 328, 339–40
  Piedrahita and, 314–16
Alzforum, 197
Alzheimer, Alois, 19–21, 50, 275
Alzheimer's Association, 37, 316
Alzheimer's Association International Conferences
  2018, 171–72
  2019, 233–34, 236–41
  2020, 289–90
  2022, 333, 335–41
Alzheimer's & Dementia, 169, 170
Alzheimer's disease
  biochemistry. See biochemistry
  clinical trials. See clinical trials
  definition of, 21–22
  genetics. See genes and genetics
  history, 19–24
  symptoms, 8–9, 26, 28, 47, 55–56, 68–69, 70, 75–76, 82, 94, 125, 144–45, 160, 172, 200, 201, 223, 261, 275–76, 320
Alzheimer's Disease Awareness Month, 37
Alzheimer's disease vaccine, 34, 81–85, 100–101, 346
  Schenk's, 81–82, 83, 90, 100
Alzheimer's Prevention Initiative trial. See API Colombia drug trial

INDEX

"amnesia," 218
Amparo, 162–67, 279
amyloid-beta, 22–23, 31, 81–82, 107–8, 201, 295–96
amyloid-beta precursor protein (APP), 22, 38, 47, 53, 81
amyloid-beta 42 (Aß42), 64, 172
amyloid hypothesis, 22–23, 64, 85, 102, 104, 113, 172, 233–35, 242, 275, 316, 325–27, 335, 343, 344, 345
Angostura, 39, 46, 130–31, 174, 177–79
Aníbal, 113
año rural, 16
antidepressants, 187, 237
Antioquia
  history of, 6–7, 77, 177
  paisa mutation, 11–13. *See also* E280A
API Colombia drug trial, 88–91, 96–115
  Aliria case. *See* Piedrahita de Villegas, Aliria
  amyloid hypothesis, 102, 104, 113, 172, 325–27, 343, 344, 345
  author's initial meeting of families participating, 111–15
  background and origins, 86–91
  brain imaging, 98–99, 102
  Christmas parties, 111–15, 181–83, 268–69
  compensation, 89, 104, 248
  completion, 331–33, 336–38, 343–44
  control subjects, 69, 217
  COVID-19 and, 286–87
  C1 families (Belmira), 46–47
  C2 families (Canoas), 46–47, 65, 76–77, 173, 215–16, 243–44, 244, 251
  C3 families (Yarumal), 46–47
  dosage, 101–3, 107–8, 111–12, 114, 172, 183, 215, 286, 287, 325
  drug candidates, 100–103
  El Tiempo reporting, 259–60, 267, 303
  ethics and cultural sensitivities committee, 99–100, 170–71, 238
  families and recruitment, 96–97, 101–5, 107, 111–15, 141, 143–51, 215–29, 268–69
  genetic counseling and disclosure, 99–100, 115, 146, 153–54, 213
  Karlawish and, 99–100, 237–39, 337–38
  naming, 104–5
  New York Times reporting, 97, 98, 257–60, 332
  open-label extension, 236, 238–39
  other clinical trial failures and, 195–96, 197, 200–201, 206, 253–54
  public announcement of, 68, 97–98, 100
  publishing and authorship question, 169–71
  Roche management of, 108
  safety concerns, 100–101, 181
  60 Minutes reporting, 107, 113
  social support programs, 162, 286–87, 292–93
  trial design, 99–100, 237–38
APOE (apolipoprotein E), 53, 82, 86, 243, 257–59, 276, 299, 326
APOE2, 257–58
APOE3, 257–58
APOE3 Christchurch mutation, 257–62, 268–69, 276, 278, 289, 299–300, 308–9, 315, 327–28, 333, 344, 363–64*n*
APOE3-Jax, 315–16
APOE4, 82, 86, 149, 233–34, 257–58, 276, 315–16, 316, 333
APP (amyloid-beta precursor protein), 22, 38, 47, 53, 81
Arango, Juan Carlos, 48–50, 63–64
Arboleda-Velasquez, Joseph, 240–41, 244, 295, 322–23
  Aliria case, 257, 260–61
arboloco, 95–96
Arcos-Burgos, Mauricio, 46–48, 73–74, 76, 78
arrieros, 176
astrocytes, 123, 126, 299–300, 344–45
Atlético Nacional, 14, 111
autism, 23, 24, 334
autosomal dominant mutation, 9–10, 47, 57, 194
axons, 240
Ayora, Margarita, 75–76, 79

bacterial meningitis, 33
Baena, Ana, 298
Baltimore, David, 196
Banner Alzheimer's Institute, 87, 345
  API Colombia drug trial, 87–91, 96–116, 144, 292. *See also* API Colombia drug trial
  completion, 331–33
  consulting fees, 292
  press conference, 331–32
  umibecestat trial, cancellation of, 233–34, 241–42

INDEX

Barrio Pablo Escobar, 261–62, 296–99, 302, 327
basuco, 25, 33
Bateman, Randall, 143–48
   DIAN trial, 144–48, 150, 153–54, 185, 210, 213, 234–35
   genes and ancestry, 150
   genetic disclosure and counseling, 153–54, 210
   Girardota family, 145–48, 213
Bedoya, Gabriel, 211–14, 288
Begley, Sharon, 233
Belluck, Pam, 97, 257, 332
Belmira, 3–4, 5, 8, 18, 27, 39, 46–47, 65, 216, 217, 281–82
Beowulf, 242
beta-amyloid. *See* amyloid-beta
beta-amyloid hypothesis. *See* amyloid hypothesis
Betancur, Leonardo, 17
Bienestarina, 131–32
biochemistry, 22–23, 30–32, 53–54, 57–59, 81–83, 107–8, 201, 257–62, 275–77, 295–96, 344–45
   amyloid hypothesis. *See* amyloid hypothesis
   cholesterol, 82–83, 86, 147, 259
   inflammation, 123, 276, 315
   tau, 31–32, 86, 144, 172, 234, 237, 243, 258, 268–69, 275–77, 344–45
Biogen, 326, 337
   aducanumab clinical trial, 200–201, 233, 253–54, 267, 315
   aducanumab FDA approval, 316–18, 325
   Aduhelm, 317–18, 325, 334–35, 337
birth control, 88, 99, 147–48
bitter rue, 156
Black people
   African descent and genetics, 147–51, 185, 189–91, 236–37, 360–61n
   Girardota family, 145–54
Bogotá, 7, 32–34, 138, 147–48, 155, 287–88, 311
brain banks, 12. *See also* Neurociencias brain bank
brain imaging, 7–8, 83, 86, 91, 97, 98–99, 159, 240–41
Brussels, 18, 23
Builes, Miguel Ángel, 7, 177
buñuelos, 283

Cadavid, Liliana, 76–79, 359n
Cali, 144–45, 311
Calle, Julián, 69–71
Canoas, 39–46, 46–47, 56–57, 93, 96, 174
cardiac pacemakers, 34–35
Carmen, 66, 67
Carrillo, Maria, 234, 336
Cauca River, 216, 282
cerebral toxoplasmosis, 33
Checho, 161, 198
Che Guevara, 176
chivas, 40, 41, 45, 174
cholesterol, 82–83, 86, 147, 259
Chomsky, Noam, 54
Christmas, 111–15, 181–83, 235, 268–69, 338
chromosomes, 52
chromosome 4, 52
chromosome 14, 47, 53, 57, 58
chromosome 19, 53
chromosome 21, 22, 53
   APP, 22, 38, 47, 53, 81
Ciencia con Alma, 112
Cifuentes, Rodrigo, 178–79
clinical trials
   API Colombia. *See* API Colombia drug trial
   Biogen aducanumab, 200–201, 233, 253–54, 267, 315
   cancellations of, 195–96, 197, 200, 206, 233, 253–54
   DIAN, 144–48, 150, 154, 185, 207, 210, 213, 234–37
   umibecestat, 233–34, 241–42
cocaine, 15–16, 25, 217, 282
cognitive tests, 10, 86, 88, 91, 151, 156, 255, 262, 332–33
Colombia, history, 6–7, 15–18, 23–26, 79, 177–78, 217–18, 281–82
Colombian Constitution of 1991, 38
Colombian Institute of Family Welfare, 219–20, 222, 224, 227, 252
Colombian mafia, 15, 48, 97, 134, 217–18, 219, 288
comunas, 13, 25, 55, 79, 103
Comuna 13, 132–35, 140, 157, 311, 323, 329
conferences. *See* Alzheimer's Association International Conferences
consulting fees, 292, 336–37
Cornejo, William, 14–15, 18, 22
   Pedro Julio Pulgarín case, 4–5, 8–10, 15, 22

INDEX

coronary artery disease, 82–83
COVID-19 masks, 287–90, 300
COVID-19 pandemic, 269, 281, 290–91, 314
   quarantines, 285–87, 289, 311
COVID-19 vaccines, 311, 313
crenezumab
   access concerns, 238–39, 279, 317, 318
   cancellation of trials, 195–96, 206, 233–34, 253
   clinical trials, 11–12, 101–5, 107–15, 181, 195–96, 206–7, 233–35, 268–69, 316, 331–32. *See also* API Colombia drug trial
   dosage, 101–3, 107, 108, 111–12, 114, 172, 183, 215, 286, 287, 325
   research publication, 169–71
   side effects, 113, 197
CT (computed tomography) scans, 7–8, 26
cyclotron, 102

dementia tests, 9, 26, 86, 91, 156
desmentizado, 188
Deter, Auguste, 19–21
diapers, 89, 120, 137, 138, 140, 167, 168, 184, 198, 201–5, 238, 239, 248
DNA sequencing, 31–32, 47
Dominantly Inherited Alzheimer Network (DIAN), 144–48, 150, 154, 185, 207, 210, 213, 234–37
Doña Blanca, 109
donanemab, 316
Doody, Rachelle, 238–39, 332
Down syndrome, 22
drug trials. *See* clinical trials
Dumar, 307, 308, 314

early-onset Alzheimer's disease, 10–11, 21–22, 37–38
   API Colombia trial. *See* API Colombia drug trial
   APP and chromosome 21, 22, 38, 47, 53, 81, 333
   genetic counseling and, 153–54
   Pulgarín case, 3–10, 15
EEG (electroencephalography), 7–8, 240–41
Elan Pharmaceuticals, 81–82
elderflowers, 156
electrophoresis, 30–31, 47
Elida, 226, 248, 252–53, 255, 320
   death of, 289–91, 340

El Tiempo, 259–60, 267, 303
el viejito, 302, 306, 328
enyerbado, 27–28, 42, 222. *See also* witchcraft
Escobar, Pablo, 14, 15, 25, 36, 48–49, 55, 111, 240, 261–62
euthanasia, 321
exorcisms, 24
E280A (paisa mutation), 61–71, 83, 87, 93–94, 207–8, 215–29, 238
   API Colombia trial, 88, 96–116, 149–50. *See also* API Colombia drug trial
   origins of, 73–80

Fabiola, Rosa, 51
Facebook, 223, 224, 247, 248, 249, 312, 322
FARC (Revolutionary Armed Forces of Colombia), 17–18, 41, 97, 105, 177–78
Fernay, 157, 158, 159, 272, 273, 277, 288, 312
fertility rates, 13
Florelba, 48–50, 63
Food and Drug Administration (FDA), 316–18
formalin, 48, 49, 207–8
Frankfurt Hospital for the Mentally Ill and Epileptics, 19–20
Freud, Sigmund, 19–20
frontotemporal dementia, 237

Gambia, 151, 189–90
gamines, 33
gamma-secretase modulators (GSMs), 345–46
gantenerumab, 234–35, 267, 316
García Márquez, Gabriel, 11, 51, 80, 265
genealogies, 9–10, 39, 46–47, 52, 70, 96–97
   linkage studies, 52–54
   López Pineda family, 159–60, 173–80, 221
   paisa families, 39–40, 44–45, 47–48, 51, 54, 56, 62–63, 75, 114, 188
   Pulgarín family, 8–10, 18, 22, 27, 65, 216–17, 220–29
gene expression, 207
Genentech, 101–3, 105, 108, 336

INDEX

genes and genetics, 8–10, 22–23, 37–38, 46–47, 52–58, 257–62, 315–16
  African descent and, 147–51, 185, 189–91, 236–37, 360–61*n*
  amyloid hypothesis. *See* amyloid hypothesis
  APOE. *See* APOE
  E280A. *See* E280A
  Indigenous Colombians and, 6, 42, 149–50, 236–37
  I414T mutation, 189–90
  I416T mutation, 145, 150–51
  1000 Genomes Project, 150–51
genetic counseling, 67, 68, 100, 125, 146, 153–54, 183, 210–13, 319, 340
genetic disclosure, 66–67, 68, 99–100, 113, 115, 146, 182–83, 206, 210–14, 267–68, 355*n*
genetic isolate, 149–50, 185
genetic linkage studies, 52–54
genetic resilience, 344
genetic testing, 66–67, 99–100
  Daniela López Pineda, 140, 157–59, 160–61, 193–96, 265–66
genograms, 23, 46–48
  Piedrahita family, 43–45
  Pulgarín family, 9
"ghost" cells, 50
Gilberto, 304–9, 314
Giraldo, Margarita, 74–75, 219–20, 254–56
Girardota (town), 185–86, 189–90, 263
Girardota family, 145–54, 213, 262
Girardota mutation, 145–54, 182, 185–91
Glenner, George, 22
Gloria, 160–61, 198, 199
glucose, 86, 144
glutamic acid, 61
Goate, Alison, 53–54, 57–59, 73, 79, 235–36, 261
Gracia, Alta, 51
Grupo de Neurociencias de Antioquia, 10–15, 345
  Aguillón named director, 345
  brain bank. *See* Neurociencias brain bank
  Christmas parties, 111–15, 181–83
  new campus, 83–84
  paisa mutation, 10–15

hallucinations, 70, 128, 156
haplotype, 57–58, 78, 149–51
Harvard Medical School, 30, 32, 35, 36, 50, 62, 63–64, 85, 172, 240–41
Hayworth, Rita, 37
headaches, 4, 27, 68–69, 127, 272
heart attacks, 82–83, 130–31
herbs (herbalism), 42–43, 75–76, 95–96, 155–56, 281
Herencia, 365*n*
heritability, 9–10, 31, 47, 52, 55–58, 70, 93, 211
Ho, Carole, 101–3, 108, 203–4
Holtzman, David, 276
Holy Week, 186, 218
homozygotes, 70–71, 93–94
Hospital Mental de Antioquia, 69–70, 134, 223
Hospital Para Locos, 69
Hospital San Juan de Dios, 33–34
Hotel Los Recuerdos, 174, 176, 177
Hotel Tequendama, 33
Huang, Yadong, 333
Huntington's disease, 31, 52, 66, 80, 89, 122

immunohistochemistry, 30–31, 61, 124
Indigenous Colombians, 6, 42, 149–50, 236–37
infant mortality, 16
inflammatory hypothesis, 123, 276, 315
InterContinental Hotel, 35, 143, 149
International Hemp Solutions, 266
in vitro fertilization, 236

José María Córdova International Airport, 98, 247, 249, 256, 290
Journal of the American Medical Association, 68

Karlawish, Jason, 99–100, 237–39, 337–38
Katzman, Robert, 37
Kosik, Ken, 29–36, 51–59
  Aliria case, 244, 258, 260, 261, 276, 299–302, 323, 326–27, 333, 339–40
  Alzheimer's Association Conferences, 236–37, 336
  Alzheimer's vaccine, 83
  API Colombia trial, 88–91, 100–104, 109–10, 143–45, 154, 206–10, 266–67, 343–44
  background of, 29–32
  Banner and, 87, 88

INDEX

Kosik, Ken *(cont.)*
  brain bank and, 62, 206–10, 299–302, 322
  crenezumab trials cancellations, 195–96
  at Harvard, 32, 35, 50, 63–64, 85, 119, 240
  initial Colombia visit, 32–36
  Lopera and, 29–30, 36, 38–39, 47, 51, 54–56, 58–59, 62, 65–67, 68–69, 110, 169–71, 188–89, 261, 336
  at McLean Hospital, 30–31, 62
  paisa genealogies, 51, 54, 56–57, 70–71, 87
  paisa mutation (E280A), 38–39, 47, 50, 53–56, 62, 65–71, 78, 86–87, 101–2, 149–51, 188–89, 207–9, 260, 262
  presenilin mutation, 58–59, 61, 85, 143–45, 147, 150, 185, 207, 236–37, 266–67, 343–44
  publishing and authorship question, 169–71
  Quiroz and, 240–41, 258, 261
  tau, 31–32, 237, 258, 275, 276–77
  at UC Santa Barbara, 85–86, 169, 336

La Ceja, 222–26, 249, 252, 289–90, 319, 338
Lacks, Henrietta, 259, 299–300
Lancet Neurology, The, 208
las familias, 12
late-onset Alzheimer's disease, 21–22, 53, 82, 207, 240, 257–58, 315, 326
"Lazarus," 253–54
lecanemab, 316, 335
Lemere, Cynthia, 63–64
Lesné, Sylvain, 335
linkage studies, 52–54
Llibre-Guerra, Jorge, 362*n*
loco, 96
London mutation, 53
Lopera, Luis Emilio, 84
Lopera Restrepo, Francisco
  Aliria case, 257, 259, 261, 267, 303, 313–14, 327, 328, 339–40
  Alzheimer's Association Conferences, 171–72, 239–40, 289, 336–37
  Alzheimer's vaccine, 83–85, 100
  amyloid hypothesis, 22–23, 113, 234–35, 325–27, 345
  API Colombia trial, 11–12, 96–116, 188–89, 196–97, 206, 238, 266–69, 292, 325–27, 336–37
  background and origins, 87–91
  completion, 331–33, 336–37, 338–39
  families and recruitment, 96–97, 101–5, 111–15, 143–51, 266–68, 328–29, 338–39
  Phoenix meeting, 90–91
  safety concerns, 100–101
  background of, 5–6, 80
  in Brussels, 18, 23
  Christmas parties, 111–15, 181–83, 268–69
  COVID-19 and, 286, 287, 289
  crenezumab trials cancellations, 253–54
  DIAN and, 144–48, 150, 213, 235
  Doralba case, 127–28, 130
  early Alzheimer's research, 10–15, 22, 24, 29, 37, 41, 46–48, 51, 58–59, 80
  genetic counseling issues, 67, 68, 153–54, 210–13, 340
  genetic disclosure issues, 66–67, 99–100, 113, 115, 146, 182–83, 206, 210–14, 267–68, 355*n*
  kidnapping of, 14–15, 356*n*
  Kosik and, 29–30, 36, 38–39, 47, 51, 54–56, 58–59, 62, 65–67, 68–69, 110, 169–71, 188–89, 261, 336
  marijuana study, 266–69
  Ofelia case, 26–28, 46, 64–65, 67, 84, 215–16
  paisa families, 37, 38–39, 41, 45–48, 49, 51, 54–59, 62–63, 67–71, 73–80, 87, 266–68, 328–29
  paisa mutation (E280A), 10–15, 24, 61–62, 66–67, 68–69, 73–80
  Pedro Julio Pulgarín case, 5, 7–10, 22, 24, 80
  presenilin mutation, 61, 147, 182, 188–89, 235
  publishing and authorship question, 169–71
  rabies case, 355*n*
  right-wing paramilitarism and, 16, 17–18
  Rodrigo case, 25–26, 27–28
López, Abel, 130–31, 174–75, 198
López, Javier, 175–76, 180
López Pineda, Daniela, 213–14, 253, 311–13
  as author's roommate, 265–66, 269–72, 291–93

INDEX

COVID-19 and, 287–88, 311
family background of, 129–35, 137–39
family genealogy, 159–60, 173–80
Fredy and, 320–21
genetic testing, 140, 157–59, 160–61, 193–96, 265–66
mother's brain autopsy, 124–28, 130, 155–56
Spain plans, 312–13, 321–22, 323, 329–30, 334
Ximena and, 138, 140, 155–57, 173, 193, 266, 272
López Pineda, Doralba, 129–41, 265, 312
background of, 130–41
brain autopsy, 118–28, 129–30, 139–40, 155
clinical history, 126, 156
death of, 139–40, 200
diagnosis of, 127–28, 156, 166–67, 346
funeral of, 140
López Pineda, Elcy, 159, 163, 166, 204–5, 270, 271
death of, 272–74, 277
López Pineda family genealogy, 159–60, 173–80, 221
López Pineda, Flaca, 162, 193–94, 203–4, 266, 269, 272, 279–80, 287–88, 312
background of, 129–30, 133, 135–41
Daniela and genetic test, 157–58
Elcy and, 205, 273
family genealogy, 159–60, 173–80
Mabilia and, 197–200, 277–79
mother's brain autopsy, 124–25
wedding of, 334, 346
López Pineda, Fredy, 159, 163, 166, 168, 270–71, 320–23
brain autopsy, 320–21, 322–23
López Pineda, Jaime, 131, 133, 162–63, 164, 205, 270, 273
López Pineda, Lina, 162–68, 271, 273–74, 277–79, 313, 334
López Pineda, Mabilia, 205
death of, 274, 277–79
impairment of, 160–61, 163, 168, 197–99, 269–70
López Pineda, María Elena, 168, 204–6, 241, 279–80
Alzheimer's vaccine, 346
Daniela and genetic test, 157–60, 195
Elcy and, 204–6, 273, 274
family genealogy, 159–60, 173–74

impairment of, 140, 141, 204, 205–6, 271, 279–80, 322
Lina and, 163–64, 168
Mabilia's death, 277–79
Lowe, Derek, 334–35
LRP1 (low-density lipoprotein receptor-related protein 1), 276

McLean Hospital, 30–31, 62
Madrigal, Lucía, 79
Doralba case, 137–38
paisa families and genealogies, 39–40, 44–45, 47–48, 51, 54, 56, 62–63, 75, 114, 188
Pulgarín case, 9
Rodrigo and Ofelia cases, 27–28
Magaly, 296–97, 298, 302–9, 313–14, 327–28
Málaga, Spain, 334
malaria, 34
Manicomio Departmental, 69
Manrique, 66
manzanillo, 95–96
Mariano de Jesús Euse Hoyos (Padre Marianito), 40, 76, 95, 105, 136, 174, 177–79
marijuana, 266–68
Masliah, Eliezer, 365–66n
masochism, 158
Matallana, Diana, 36
Medellín, 13–14
author's move to, 10–11
COVID-19 quarantine, 281, 285–87
history of, 6–7
Kosik's visit, 35–36, 38–39
murder rates, 55
Medellín Airport, 98, 247, 249, 256, 290
Medellín Cartel, 23–25, 36, 37, 48–49, 134–35, 240, 256
Medellín massacre of 1987, 16–18
Medellín River, 186, 225
Medical College of Pennsylvania, 29–30
Medicare, 317, 334
médico rural, 46
melanoma, 295, 297–300, 345
memory binding tests, 91
memory tests, 91, 156, 159, 276
Meneces, Paulina, 190–91
microglia, 172, 315
Miriam, 249–51, 252, 254–56, 319–20
missense mutation, 61–63. *See also* E280A
monoclonal antibodies, 90, 101, 328

373

INDEX

Monserrate, 33
montañero, 3, 148
Montanoa quadrangularis, 96
Monte Delphos, 109
Moreno, Sonia, 74–75, 113, 114, 151, 153, 186, 187, 189–91, 283–84
Mount Parnassus, 23
muchacha, 193
Muñeton, Esteban, 121–22, 183–84, 273, 300

National Geographic, 321
National Institute on Aging, 37
National Institutes of Health (NIH), 32, 102, 143, 207, 263, 301
National University of Colombia, 33
Nature (journal), 34, 58
Nature Genetics, 58
Nature Medicine, 257, 258
Neurociencias. *See* Grupo de Neurociencias de Antioquia
Neurociencias brain bank, 12, 62, 119–28, 183–85, 206–10
  Aliria, 298–302
  Camila, 262–64
  COVID-19 and, 288–89
  Doralba López Pineda, 119–28
  Elcy, 272–74
  Florelba, 48–50, 63
  Fredy, 320–21, 322–23
  funding, 143, 207, 208, 262–63, 301
  German and American collaboration, 206–10
  Mabilia, 277–79
  Ofelia, 26–28, 215–16
  Pedro, 7–8
neurodegeneration, 172, 200
neurofibrillary tangles, 20–21, 31, 50–51, 275–76
neurofilament light chain (NfL), 240, 243
New York Times, 97, 98, 257–60, 332
niacin, 4
Nobel Prize, 196
Novartis, 83–85

Ofelia, 26–28, 46, 64–65, 67, 84, 215–16
One Hundred Years of Solitude (García Márquez), 11, 51, 80, 265
1000 Genomes Project, 150–51
open-label extension, 236, 238–39
Open Philanthropy, 295–96
Orlando, 132–40, 177

Osorio, Enrique, 32–33, 36
Ossa, Jorge, 38–39, 54, 56, 57, 73

pacemakers, 34–35
Padre Marianito, 40, 76, 95, 105, 136, 174, 177–79
paisa mutation. *See* E280A
paisas, 6–7, 24–25
paracos, 135
Paraíso, 263
Parkinson's disease, 80, 122
Pasto, 266
Patarroyo, Manuel Elkin, 34–35, 39
patents, 244, 258, 295, 303, 328
"patient zero," 78
Paula, 221, 222, 223–29, 247–56, 252–53, 289–91, 319–22, 333, 340
  genetic testing, 319–20, 322
"Peculiar Disease of the Cerebral Cortex" (Alzheimer), 20–21
Pedregal, 131, 160, 163, 270, 273
Pentecostalism, 136–37
Petro, Gustavo, 155
Petryna, Adriana, *When Experiments Travel*, 362–63n
PET (positron emission tomography) scans, 98–99, 102
phantom limbs, 24
Piedrahita de Villegas, Aliria, 295–309, 313–16
Piedrahita, Danilo, 93–96
Piedrahita, Francisco, 40–46
  background of, 40–43
  genetic testing, 315
  Neurociencias and paisa families, 40–41, 44–46, 114, 115, 162, 196–97, 314–16, 340–41
Piedrahita, Ledy, 42, 44, 93–96, 97, 103, 307
Pineda, José Miguel, 76, 173
Pineda, Mariana, 179–80
Pineda, Roselia, 163, 173
Pineda Sampedro, Mauricio, 76–79
Pineda, Sol, 179–80
polygenic risk scores, 235–36
"possible biological efficacy," 333
Potamkin Prize, 366n
presenilin mutation, 58–59, 61, 64, 82–83, 85, 115, 143–45, 147, 150, 182, 185, 188, 207, 235, 236–37, 258, 266, 333, 343–44
proof of concept, 263

INDEX

Pulgarín Balbín, Carlos, 224, 247–48, 249, 252–53, 291
Pulgarín Balbín, Marcela, 220–29, 247–56, 289–91, 338, 340
Pulgarín Balbín, Marta, 220–29, 247–56, 289–91, 320, 333, 338–40
Pulgarín Balbín, Mery, 220–29, 247–56, 290–91, 320
Pulgarín, Blanca Nidia, 216–19, 281–85, 288, 298, 338
Pulgarín family and genealogy, 8–10, 18, 22, 27, 65, 216–17, 220–29
Pulgarín, Pedro Julio, 24, 219, 282
   death of, 18
   García Márquez and, 11, 80
   hospitalization of, 3–5

Q'hubo, 103, 114, 165, 260, 307
Quiroz, Yakeel
   Aliria case, 242, 244, 257–62, 289, 296, 308
   Alzheimer's Association Conferences, 242, 289
   brain imaging, 240–41
   E280A and APOE3 Christchurch research, 239–41, 242, 257–62, 276, 289

rabies, 244, 355n
radioactive tracers, 102
Radio Paisa, 133, 136
Ramírez, Laura, 151–53
Rauch, Jennifer, 276
"red zones," 97
Reiman, Eric
   Alzheimer's Association Conference (2019), 233–34, 237, 239–40
   API Colombia trial, 88–91, 96–97, 100, 105, 144, 154, 197, 237, 239–45, 257–59
   Aliria case, 232–45, 257–59, 297
   background and origins, 86–91
   completion, 331, 332, 337
   genetic counseling, 154, 183
   Phoenix meetings, 90–91, 100
   umibecestat trial, cancellation of, 233–34, 241–42
RELN gene, 315, 344
Respa, Gilma, 51
Restrepo, Carlos, 212–13
Revolutionary Armed Forces of Colombia (FARC), 17–18, 41, 97, 105, 177–78

Reynolds, Jorge, 34–35
Rey–Osterrieth complex figure, 26
right-wing paramilitarism, 15–18, 23–25, 41, 56–57, 77, 134–35, 281
Ríos, Silvia, 200–201
rivastigmine, 156, 361n
Roche, 102, 108, 113, 238, 316, 339, 345–46
   crenezumab trials cancellation, 195–96, 197
   Lopera's financial disclosure, 336–37
   open-label extension, 236, 238–39
Rocío, 302–3, 305–9, 314, 327, 328
Rodrigo, 25–26, 27–28, 46–47
Roldán, Mary, 190
Rua, Alberto, 189
Rua, Camila
   brain autopsy, 262–64
   end stages of, 187, 202–3
Rua, Cenen, 189
Rua, Horacio, 188, 189–90
Rua, Indalecio, 190
Rua, Manuela "Ruda," 190–91
Rua, María Mathilde, 190
Rua, Oliva, 188, 189, 249, 340
Rua, Piedad, 185–91, 201–3, 239, 262–64, 270, 288
Rua, Trinidad, 186–87, 189

sainete, 148–49
St George-Hyslop, Peter, 58
salmon knife, 122–23
Sampedro, Petronila, 76–79, 173
Sampedro sisters, 76–79
San Andrés de Cuerquia, 77
San Vicente de Paúl hospital, 13–14, 15, 24, 55, 111, 217
   Ofelia case, 26–28, 46, 64–65
   Pedro Julio Pulgarín case, 3–5, 216
   Rodrigo case, 25–26, 27–28
Schenk, Dale, 81–82, 83, 90, 100
schizophrenia, 70
Science (magazine), 334–35
scientific imperialism, 169, 210
Sean, 249, 250, 254–56, 319–20
Sebastián, 198–99, 277–79
Sede de Investigación Universitaria (SIU), 110–11, 119–20, 196–97
   Christmas parties, 111–15, 181–83, 268–69
Selkoe, Dennis, 30–32, 63
Sepulveda-Falla, Diego
   Aliria's brain, 299–302, 322–23, 344

INDEX

Sepulveda-Falla, Diego (*cont.*)
   German lab of, 124, 207–10, 208–10, 322–23
   study of E280A brains, 124, 207, 208–10, 301–2, 322–23
side effects, 113, 154, 197, 220, 345
single-cell RNA sequencing, 207–8, 262, 273
Sinú River, 236
SIU. *See* Sede de Investigación Universitaria (SIU)
60 Minutes (TV show), 107, 113
slavery, 6, 189–91, 190–91
soccer, 14, 111
solanezumab, 107–8, 114–15, 234–35
Spain, 312–13, 321–22, 323, 329–30, 334
Spanish conquest, 6, 10, 42, 78, 150
speech disorders, 54
statins, 82–83, 147, 148
sterilization, 212–13
Strauss, Richard, 242–43
suicide, 67, 126, 141, 168, 183, 321, 340
synthetic amyloid-beta, 81–82
syphilis, 7, 33

Tariot, Pierre, 87
   Alzheimer's Association Conference, 233–34, 237
   API Colombia trial, 88–91, 96, 98, 100–101, 105, 108–9, 144, 241–45
      background and origins, 88–91
      completion, 331–32, 337
      genetic counseling, 183
      Phoenix meeting, 90–91
   background of, 87
   umibecestat trial, cancellation of, 233–34, 241–42
tau, 31–32, 86, 144, 172, 234, 237, 243, 258, 268–69, 275–77, 315, 344–45
temporal lobes, 7
Tessier-Lavigne, Marc, 365*n*
"translational gorilla," 85
traumatic events and dementia, 135
Trump, Donald, 208
Tufts University, 30
tullido, 188
Twitter, 333

umibecestat, 233–34, 241–42
University of Antioquia, 38, 83–84, 209–10
   massacre of 1987, 16–18
   San Vicente hospital. *See* San Vicente de Paúl hospital
   Sede de Investigación Universitaria. *See* Sede de Investigación Universitaria
University of California, Santa Barbara, 85–86, 169
University of Minnesota, 335
University of Pennsylvania, 99–100, 237–38
Uribe, Claramónika, 109, 254, 329

vaccines
   Alzheimer's disease, 34, 81–85, 100–101, 114, 346
   COVID-19, 311, 313
Villa Aliria, 327, 328, 339–40, 365*n*
Villegas, Andrés
   COVID-19 diagnosis, 288
   Daniela and, 124–27, 129–30, 141, 194, 313
   Neurociencias brain bank, 119–28, 184, 208–10
   Aliria's autopsy, 299–302
   Camila's autopsy, 263
   Doralba's autopsy, 119–28, 129–30, 155
Virgen del Carmen, 111, 218, 222

wakes, 62–63, 120, 290
Washington University, 53, 144, 145–46, 210, 261, 276
Wexler, Nancy, 31, 52, 89
WhatsApp, 119, 129, 155, 183, 207, 262, 272, 273, 282, 299, 332–33
witchcraft, 28, 42–43, 75, 177, 222–23
Wong, Caine, 22
World Health Organization, 34

Ximena, 138, 140, 155–57, 173, 193, 266, 272

Yarumal, 39, 44, 46–47, 56–57, 220–21, 266–68
Yenny, 221, 225–26, 252, 255
Yunis, Emilio, 361*n*